510 WIL

MATHS

COURSEWORK COMPANION

Ray Williams

GCSE

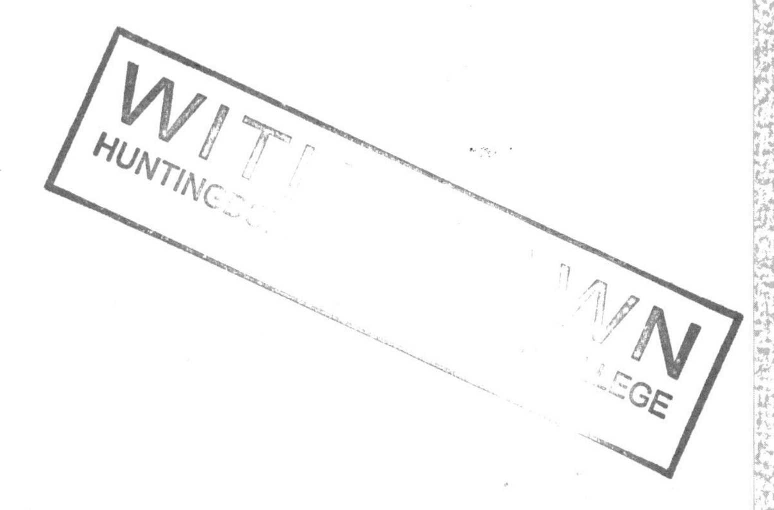

100538
THE HUNTINGDONSHIRE COLLEGE LIBRARY

Charles Letts & Co Ltd
London, Edinburgh & New York

First published 1989
by Charles Letts & Co Ltd
Diary House, Borough Road, London SE1 1DW

Text: © Ray Williams 1989
Cover photograph: Michel Tcherevkoff, Image Bank
Handwriting samples: Artistic License
Diagrams: Kevin Jones Associates
Cartoons: Viv Quillin
Illustrations: © Charles Letts & Co Ltd 1989

All our Rights Reserved. No part of this
publication may be reproduced, stored in a
retrieval system, or transmitted, in any form
or by any means, electronic, mechanical,
photocopying, recording or otherwise, without the
prior permission of Charles Letts Publishers.

'Letts' is a registered trademark of
Charles Letts & Co Ltd

British Library Cataloguing in Publication Data
Williams, Ray
GCSE mathematics
1. Mathematics
I. Title
510

ISBN 0 85097 862 9

The author and publishers would like to thank the
following for permission to reproduce copyright material:
LEAG, for 'Passageways', p. 22.
MEG, for their guidelines used on p. 57.
SEG, for 'Everything stops for tea', p. 62.

Printed and bound in Great Britain by
Charles Letts (Scotland) Ltd

Contents

THE HUNTINGDONSHIRE COLLEGE LIBRARY

Introduction

What is coursework?

At some stage during your mathematics course, you will be asked to produce pieces of work that are different in nature from the straightforward 'question and answer' techniques of most mathematics textbooks or examination papers. This different type of work is loosely described as 'coursework'. However, before you become *involved* in coursework, you should have some understanding of what it is all about.

Coursework is *not* a set of examples or an exercise done from a textbook. What is more likely to happen is that your teacher will break away from his or her normal teaching and allow you time to study some topic and then to write up a report.

Examinations test you on a particular day at the end of a course. Some people are better than others at sitting exams; we can all have 'off-days'. Coursework enables you to collect marks over a period of time *throughout* the course. You work in a completely different style. There are rarely any set answers in coursework. Your teacher is more interested in *how* you tackle the task than in what answers you arrive at.

The **national criteria for mathematics** state six **elements** for coursework. You are expected to be able to:

1 respond orally to mathematics questions
2 discuss mathematical ideas
3 perform mental calculations
4 carry out practical work
5 carry out work of an investigational nature
6 produce extended pieces of work.

All these areas should be covered within your coursework at some stage.

Although this book is divided into sections that relate to these areas, you should always remember that any given piece of coursework may contain more than one of the six elements, but check with your teacher that you have covered all of them by the time you have finished all your coursework assignments.

Remember, the GCSE examination is designed to test what you *do* know and understand, rather than what you do not know or cannot manage. You are encouraged to be more positive in your approach, since one of the main aims is to find out what you can get right, not what you can get wrong. It should be the same with all your coursework. You should feel encouraged to show what you know and you should show how your skills are developing in areas that are of particular interest to you.

You will have the opportunity to spend some considerable time during your mathematics course engaged in aspects which appeal to you. This time is very valuable to you, so *TRY NOT TO WASTE IT!*

SECTION ONE

Variations in coursework

Which group?

Your school can choose from six different examination groups as to which GCSE in mathematics you will take. This will be decided, in the majority of cases, by where your school is situated. The different groups and their addresses are given at the end of this book, in the Appendix.

It does not necessarily follow, however, that a school, say in the Midlands, will take part in the GCSE of the Midlands Examining Group. So, one of the first things you should check, is *which* examination group your school has chosen.

All examination groups have decided upon a coursework component which fulfils the requirements of the aims and objectives of the **national criteria**. The problem is that there is a variation in style of this coursework amongst the groups. In fact, each of the six groups has a different requirement, both in quantity and variety.

It is worth studying the Appendix, at the end of this book, very carefully to see the differences and to find out what is expected of you once you know which of the groups you are following.

As you will see from the Appendix, each group has different coursework sections, but there is also a great deal of overlap amongst the components.

Some of the different words used to describe the same thing could, perhaps, at first confuse you. For example, an *assignment*, a *task* or an *investigation*, could refer to the same piece of coursework. An *everyday application of mathematics*, an *extended piece of work* or a *statistical survey* can often be referred to as a *topic* or a *project*.

Getting started

The teacher's role

Once your teacher has explained to you about your assignment, you will feel the need to discuss your work with someone at some stages. Your teacher is the best person to give advice and, generally, will always be available. Even if your teacher can't help you directly, he or she should be able to suggest where you can obtain the necessary help or information. If you are working in a classroom, your teacher will probably move from group to group, so be *prepared* to ask for advice when it is your turn. Keep a notepad handy so that you can jot down the things that you would like to ask. Then, when you have those vital meetings with your teacher, time is not wasted.

Your teacher should make sure that you are keeping to your timetable and that you will finish on time. Take care with your presentation, remembering that your teacher is the one who has to mark it. Your teacher's role should be one of support and encouragement even though he or she will be the **examiner** in the end.

Forms of moderation

You might have asked yourself how your mark can be compared, fairly, with that of someone at another school when your work is marked by different teachers. Comparability is achieved by what is called **moderation**.

Moderation can take place in different ways:

- An examination group member may re-mark a certain number of assignments from your school and, if there is a disagreement, suggest that all of the school's marks are adjusted.
- Teachers are trained by the examination group to act as **moderators** who will then re-mark **sample assignments** from schools within their area.
- Groups of teachers will meet to re-mark samples of work and come to an agreement on a **standard set** of marks.

Useful hints

Whatever assignment you are tackling, don't rush straight into it. Refer to your teacher *at all times*. Check with this book at the appropriate sections.

Below is a list of *do's* and *don'ts* to give you guidance through an assignment.

Do's

- Make sure there is plenty of maths in your work.
- Try to be original.
- Use a calculator but explain your answers.
- Ask your teacher if your title and plan are suitable.
- Ask your teacher what he or she is looking for to give good marks.
- Check to see what resources are available e.g. computer, library etc.
- Make sure that you can explain everything that you have written down. Remember that your teacher has to check that your work is your own and not a copy.

Don'ts

- Don't put off till tomorrow what you can do today.
- Don't copy exactly what is on your calculator display if it gives an answer to five decimal places. Accuracy must be to a *reasonable* order of accuracy e.g. 1 d.p. or 2 d.p.
- Don't copy out of books or magazines.
- Don't paste in material from magazines or books without *explaining* what it means in relation to your work.

Checklist of questions

You should ask your teacher:

1 Which syllabus am I taking?
2 How many assignments do I have to complete?
3 Are they all equally important?
4 When do I have to hand in the assignments?
5 Will I be allowed to work on the assignments outside normal class-time?

What you must now decide is which of the Sections of this book are relevant to your particular needs. For the majority of you, this book will then provide useful material and advice and will help you to improve your GCSE coursework.

SECTION TWO

Problem solving tasks and investigations

Tackling the problem

Problems can be solved in different ways. Although the title 'Problem solving' suggests that your final solution is all-important, far greater emphasis is placed on *the way* you tackle the problem than on your arrival at a correct answer at the end. You will have to explain which method and route you are attempting, every step of the way. The title 'Problem solving' also suggests that you are not going to get a lot of help from your teacher. He or she might suggest avenues to follow if you are getting nowhere, but it may be that you will do this assignment under test conditions; then no help can be given whatsoever.

Here are the main areas in which you can demonstrate to your teacher how you would solve the problem.

Identification of the task

You have to show that you understand what the problem involves. I suggest that you read the question more than once, checking the mathematical language used. Make sure you *understand* what is meant by such words and phrases as: regular polygons, tessellation, perimeter, area etc. If the problem appears vague without too much information, your teacher is looking for you to form suitable problems from what you are given.

Planning

In this section you have to choose an overall strategy. Various strategies are explained as we tackle **Problem 1**, on page 11, together.

Try to consider a **range of methods** – there may be more than one way to solve the problem. Your teacher is looking to see if you have chosen the most suitable method. If you are doing page upon page of calculations, then you can be sure that there is an alternative and simpler method.

Try to use all of the information that you are given in planning your solution and in starting at the most appropriate point.

Finally, in this planning stage, consider whether you need equipment or additional material. Choosing the correct equipment may be an area for which marks are being awarded.

Carrying out the task

Consider these aspects of your work when you are carrying out the task.

- **Accuracy** – Were you accurate in your work and did you work to the correct degree of accuracy?
  e.g. angle $A = 36.5°$ (to 1 decimal place)
  rather than
  angle $A = 36.47862°$
  Always *check* your calculations.
- **Alterations** – Did you try a different approach in the light of the

results you obtained? You must be aware of, and take notice of, your findings.

- **Patterns** – Always look for patterns in number. Don't just look for odds and evens. Look for square numbers, Fibonacci numbers, triangular numbers etc.
- **Predictions** – Can you make predictions as to what will happen next? You must test your predictions whenever possible.
- **Generalization** – see page 13 to give you some idea of generalizations. Try to generalize if you see patterns developing. Generalize both in words and algebra if at all possible.

  Always test your generalization.

Communication

It is important that you explain your actions clearly at each stage. You must show all your recordings, preferably in a table or chart, then present your results in an appropriate way by using graphs, charts etc.

Finally, try to assess your results and see if any of them can apply to other situations. If it is possible, try to talk through your work with a friend, explaining any assumptions that you made and any pitfalls that were met.

'Even if you think a point is obvious you must still state it. Assume that your teacher needs everything explaining to him.'

As well as thinking of the four main areas of:

1 understanding **2** planning **3** carrying out the task
4 communication

also write down everything that you think is worthwhile. There may be discretionary marks for other points worthy of note.

Checklist

Here is a checklist to help you through the problems that are set at the end of this section.

1. **Understanding the task** – Check the language
2. **Planning your work** – Choosing method, information and equipment
3. **Carrying out the task** – Accuracy and strategy
4. **Communication** – Written and verbal
   Be clear and precise. Use diagrams wherever possible
   Evaluate the results

Key strategies

It would appear that the best method to tackle a problem is to arm yourself with a **checklist** that contains **key strategies**. ('Strategy' is a word used to describe a method of approach.) These are the key strategies.

(a) Try some simple cases
(b) Organize your work systematically – diagrams help
(c) Make up a table of results
(d) Look for patterns and relationships in number
(e) Use these patterns to extend the table of results
(f) Find a general rule and explain it symbolically
(g) Explain your rule. Show what you are doing and why
(h) Check to see if your general rule works

As you work through this section you will face unfamiliar problems. It should be your aim to gain experience and confidence until you realize that problem solving is made easier if you follow these simple key strategies. As you tackle a problem keep them in mind as part of your **checklist**.

An excellent problem that highlights all of these strategies is given next. Try to solve this problem about matches, then study carefully Problem 1 about office telephones, which highlights all the strategies mentioned above.

Matches

Matches are joined end to end to form various shapes. Any number of matches can be used.

Examples

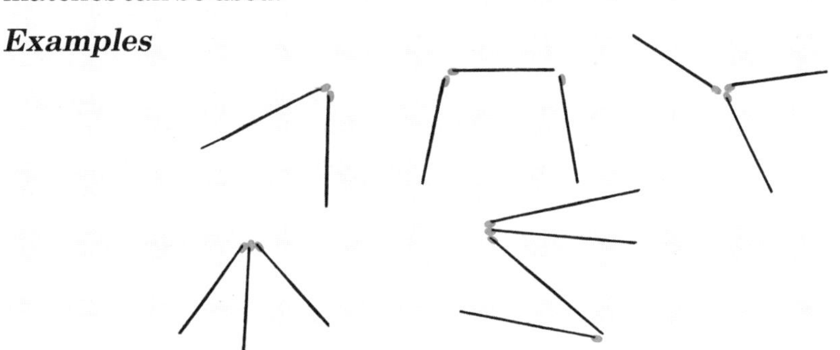

Can you find a connection between the number of matches (M), the number of joins (J) and the number of ends (E) which exist in all possible patterns?

Note closed loops are not allowed.

How many matches will be needed for a pattern which has two joins and five ends?
Can you draw a possible pattern for this?

Problem 1 Office telephones

There are 100 offices in a building – each with its own telephone. Each telephone is connected to all the other phones in the building individually. How many connections are needed so that any office can contact any other office in the building directly?

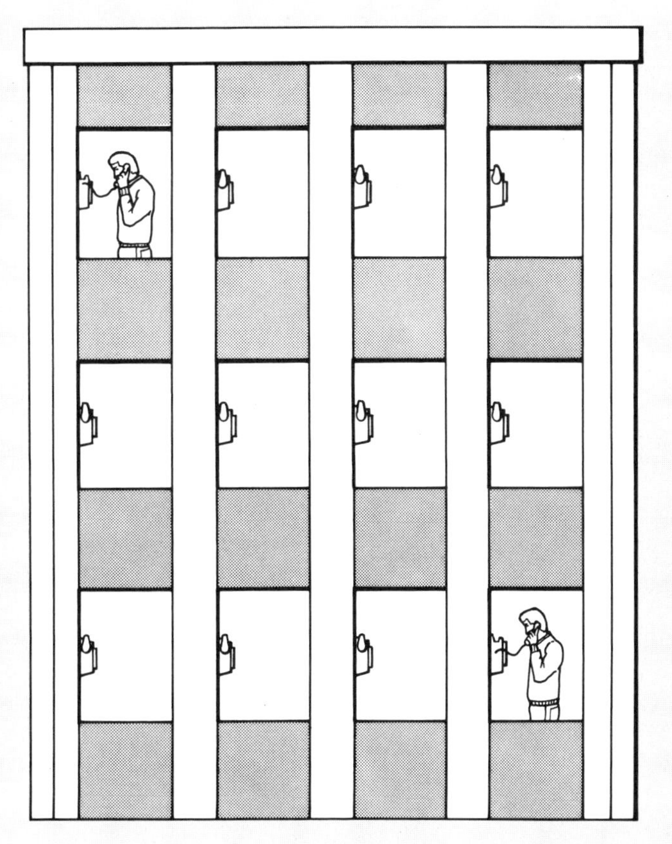

This problem cannot be solved immediately, so work through the list of key strategies on page 10.

Strategy (a) Try some simple cases

Instead of considering 100 telephones, consider three phones interconnected, then four phones, and so on.

Strategy (b) Organize your work systematically

Make a list of all the connections when using three telephones, then four phones, and so on. Show them in a diagram. Consider each phone in turn as a starting point; do not jump from one to another in a haphazard way.

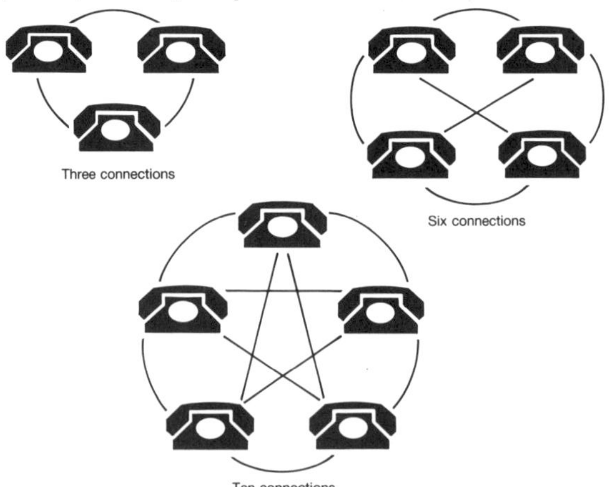

Strategy (c) Make a table of results

Number of telephones	3	4	5	6	7	8
Number of connecting lines	3	6	10			

Strategy (d) Look for patterns and relationships

The table above shows how the number of connections increases as the number of phones increases. Can you see a relationship between the two sets of numbers? Keep going until you can *predict*, that is forecast, the next number in the 'Number of connecting lines' row.

Can you see that in the table you are adding 3, then 4, then 5, then 6 to find the next number in the connecting lines row? You are now in a position to solve the problem by **counting on** until you get to 100 telephones.

Obviously this is a long way of solving the problem. Try to find a quicker method than counting on, if you can.

Can you see that $\frac{1}{2}(3 \times 2) = 3$ $\frac{1}{2}(4 \times 3) = 6$ $\frac{1}{2}(5 \times 4) = 10$?

Then you can predict $\frac{1}{2}(6 \times 5) = 15$ will be the next number in the table.

OR

Can you see that $\frac{1}{2}(3^2 - 3) = 3$ $\frac{1}{2}(4^2 - 4) = 6$ $\frac{1}{2}(5^2 - 5) = 10$?

Then you can predict $\frac{1}{2}(6^2 - 6) = 15$ will be the next number in the table.

Strategy (e) Use the pattern to extend the table of results

Use the pattern in the numbers to solve the original problem of a building with 100 phones.

Number of connections $= \frac{1}{2}(100 \times 99) = 4950$

OR

Number of connections $= \frac{1}{2}(100^2 - 100) = 4950$

Strategy (f) Find a general rule and express it symbolically

Can you write down a formula, using algebra, that will apply to every building, no matter how many phones there are in that building?

If number of connections $= c$ and number of phones $= p$

Then, $c = \frac{p}{2}(p-1)$ *OR* $c = \frac{1}{2}(p^2 - p)$

Strategy (g) Explain your rule

Can you explain to a friend what you have to do to solve the problem? Why does the rule work?

'Each telephone is connected to 99 other phones in the building. No phone is connected to itself. If there are p telephones in the building each phone is connected to $(p-1)$ other phones and therefore has $(p-1)$ connections. But this will give the case where telephone A is connected to telephone B and telephone B is connected to telephone A by a separate connection. One connection is all we need between two phones so the total number of connections in this case will be $\frac{p}{2} \times (p-1)$.'

Strategy (h) Check to see if your general rule works

Use the formula to work out the number of connections involved in a network of six phones. Could you check your answer with a diagram?

If $c = \frac{p}{2}(p-1)$ and $p = 6$ phones, then:

$c = \frac{6}{2}(6-1) = 3 \times 5 = 15$ connections.

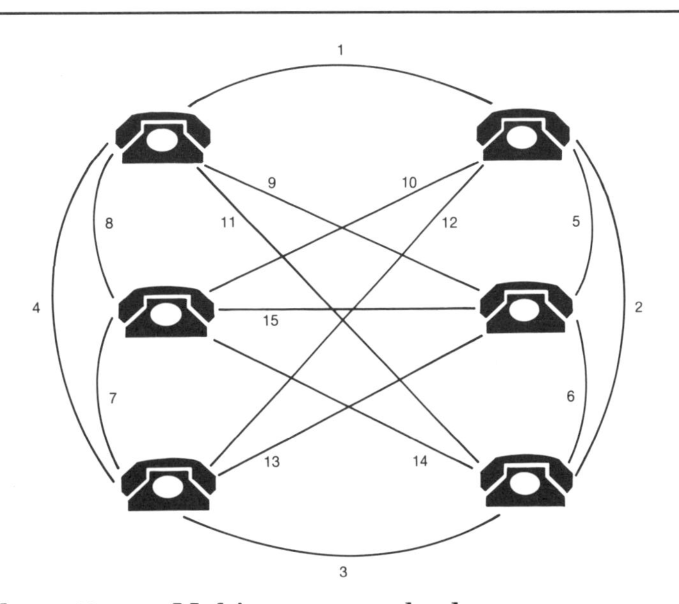

Spotting the pattern – Making a general rule

Strategies **(d)** and **(f)**, Look for patterns and relationships, and Find a general rule and express it symbolically, are the most significant parts of problem solving if you are to arrive at a solution as quickly as possible. The alternative is to go through all the stages leading up to whatever stage you are asked to solve. (Imagine what it would be like, and how long it would take you, to go through all 100 stages involved in solving Problem 1 – Office Telephones!)

Here are some useful exercises in spotting number patterns and arriving at the 'general rule'.

Example

Set A		**Set B**
1	⟶	3
2	⟶	4
3	⟶	5
5	⟶	7
8	⟶	10
x	⟶	$x+2$

We can see that each member of Set B is the corresponding member of Set A + 2.

Therefore, if we let x represent any number in Set A, then the corresponding number in Set B is $(x+2)$.

Now try these *easy* patterns and find the general rule.

Set A		**Set B**	**Set A**		**Set B**	**Set A**		**Set B**
1	⟶	5	2	⟶	7	5	⟶	3
2	⟶	6	4	⟶	9	7	⟶	5
3	⟶	7	5	⟶	10	8	⟶	6
7	⟶	11	8	⟶	13	12	⟶	10
10	⟶	??	15	⟶	??	20	⟶	??
x	⟶	( )	x	⟶	( )	x	⟶	( )

Sometimes the Set B is a multiple of Set A.

Example

Set A		**Set B**
1	⟶	2
2	⟶	4
3	⟶	6
7	⟶	14
12	⟶	24
x	⟶	$2x$

Here, each member of Set A is doubled to get the corresponding member of Set B.

Note We write $x \longrightarrow 2x$

NOT $x \longrightarrow x \times 2$

or $x \longrightarrow 2 \times x$

‘$2 \times x$ and $x \times 2$ are not incorrect but are considered clumsy ways of writing $2x$.’

Again, try these *easy* patterns and find the general rule.

Set A		**Set B**	**Set A**		**Set B**	**Set A**		**Set B**
1	⟶	3	1	⟶	5	1	⟶	$\frac{1}{2}$
2	⟶	6	2	⟶	10	2	⟶	1
4	⟶	12	3	⟶	15	4	⟶	2
5	⟶	15	6	⟶	30	8	⟶	4
10	⟶	??	12	⟶	??	20	⟶	??
x	⟶	??	x	⟶	??	x	⟶	?

Now some patterns may be a combination of multiples followed by an addition or subtraction.

Example

Set A		**Set B**
1	⟶	3
2	⟶	5
3	⟶	7
5	⟶	11
12	⟶	25
x	⟶	$2x+1$

Here, each member of Set A is doubled and 1 is added to the answer to get the corresponding member of set B.

Note how we write this using algebra:

$$x \longrightarrow 2x+1$$

The following patterns are not so easy to spot. Try to find the missing numbers and complete the general rule in each case.

Set A	1	2	4	5	10	x
Set B	5	7	11	13		

Set A	1	2	3	7	12	x
Set B	2	5	8	20		

Set A	1	2	4	5	20	x
Set B	3	8	18	23		

Set A	1	2	3	5	8	x
Set B	8	12	16	24		

If you found these last four examples difficult, here is a useful hint that works for this type.

Choose two *consecutive* numbers from Set A.

Set A	2	4	5	8	10	x
Set B	6	14	18	30		

The difference in the corresponding numbers in Set B is the multiple of x.

$18 - 14 = 4$

so $x \longrightarrow 4x - 2$ which is the general rule.

Finally, the most difficult of all the patterns to spot is when the members of the first set of numbers are squared or even cubed.

Example

Set A	3	4	5	8	10	x
Set B	6	12	20	56	90	$x^2 - x$

Notice that the hint for using consecutive numbers of Set A does not work in this case.

A most common and useful relationship to spot is the one dealing with **triangular numbers**. It would be well worth your time to study the following table carefully.

"You will be surprised to learn that of all the problems using number patterns nearly half of them involve triangular numbers. Study this table well! It may help you."

Set A	1	2	3	4	5	6	n
Set B	1	3	6	10	15	21	$\frac{1}{2}n(n+1)$

Not all problems will use the key strategies. Most problems will use some key strategies but not all.

Other strategies that you may consider when attempting this type of problem, if you cannot see where your key strategies apply, are as follows.

- **Trial and error** – Sometimes it may be possible to select a starting point at random or even select a possible solution to the problem. Apply the information that you are given to see if your guess is correct (or even close) to a solution. This may help you to see a method.
- **Process of elimination** – Once you have decided upon a starting point and you have alternative possibilities to choose from, it may be possible to reduce the alternatives by elimination.
- **Graphical representation** – It might be possible to represent the problem in pictorial form e.g. as a travel graph.

Tasks

Some examination groups use the word **tasks** to describe the assignment you have to complete. More often than not, tasks are not meant to last much longer than one hour and may be done under examination conditions.

Sample answers

Let us consider a particular problem and compare the answers submitted by two students.

Problem 2 Paving stones

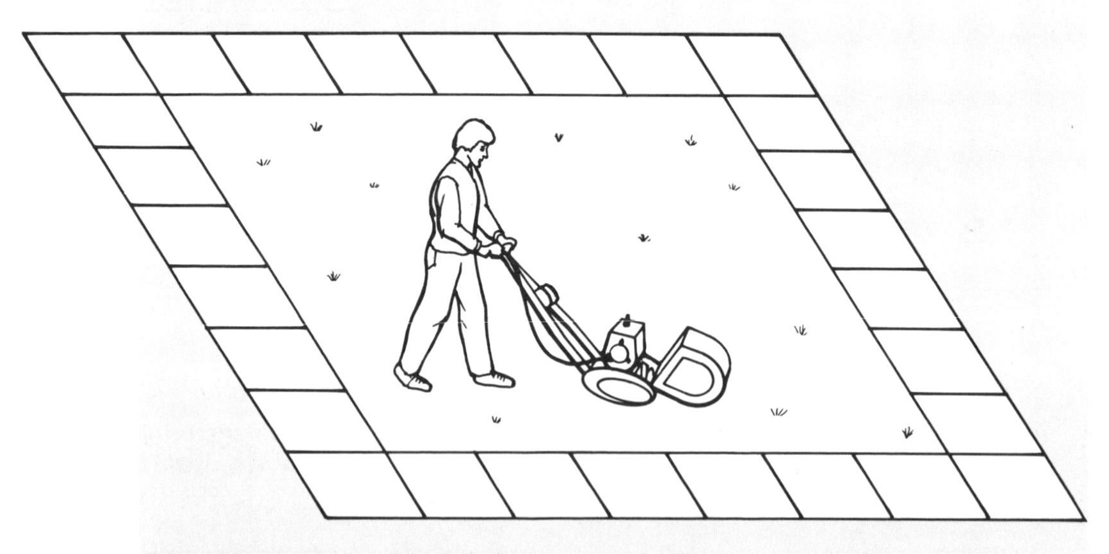

Pete works as a gardener in a stately home. In the 'formal garden' he has to lay paving stones to form paths around square lawns. The paving stones are 1 metre square.

(a) How many paving stones are needed to surround a lawn measuring 3 metres by 3 metres?

(b) How many paving stones are needed to surround a lawn measuring 4 metres by 4 metres?

(c) Find a rule that Pete can use to work out how many paving stones are needed to surround *any* square lawn.

(d) How many paving stones are needed for a 12 metres by 12 metres lawn?

(e) What are the dimensions of the square lawn where the path has 72 paving stones?

Name: Andrew

Paving Slabs

(a) 3m x 3m

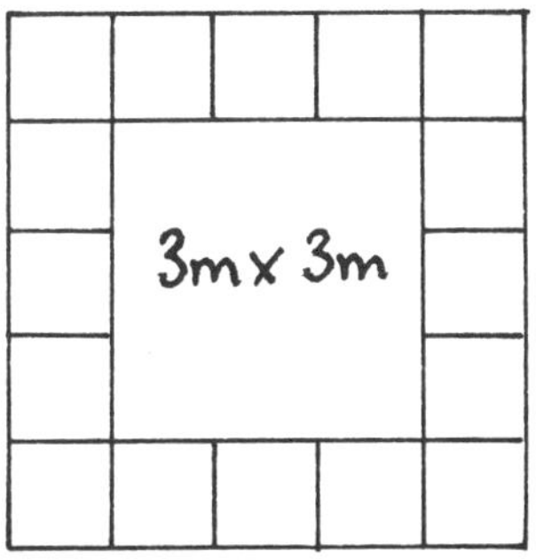

16 slabs ✓ 1 mark

(b) 4m x 4m

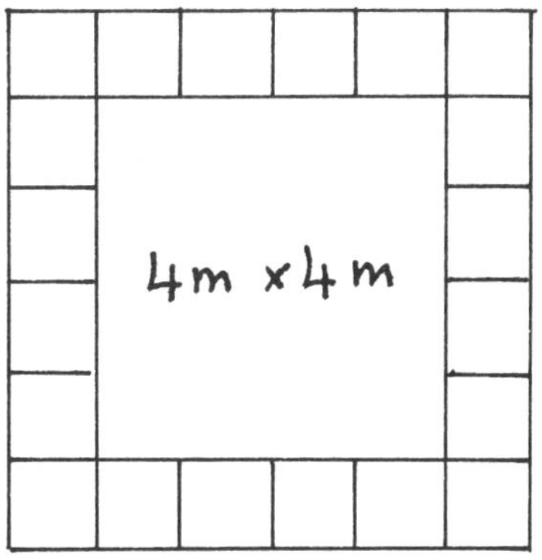

20 slabs ✓ 1 mark

Length of Lawn	No. of Slabs
3	16
4	20
5	24
6	28

✓ 1 mark

If a 6m x 6m square needs 28 slabs then

wrong basic assumption

(c) 12m x 12m = 28 x 2 = 56 slabs

Because the lawn is twice as large. It does not mean that it needs twice as many slabs.

(d) The numbers go up by 4 each time. ✓ *1 mark*

Which numbers?

(e) $\frac{72}{24} = 3$

Same wrong assumption

If a 5m x 5m lawn needs 24 paving slabs, then a 15m x 5m lawn needs 72 paving slabs.

✗

This is a poor attempt. Andrew has not referred to his checklist:

Try simple case
Be systematic
Make a table
Spot patterns
Find a rule
Check the rule

Final Mark

4/12

Name: Victoria

Paving Slabs

(a)

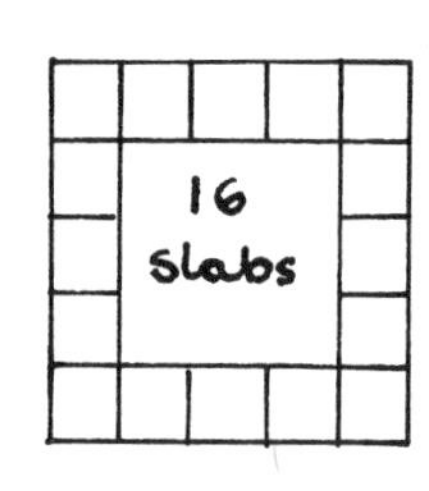

✓ 1 mark

(b)

20 slabs

✓ 1 mark

Length of lawn (n)	1	2	3	4	5	6	n
Number of slabs needed (s)	8	12	16	20	24	28	(4 × n) + 4

✓ 2 marks

(c) Number of slabs for a 12 m × 12 m lawn
= (4 × 12) + 4
= 52 slabs ✓ 2 marks

(d)

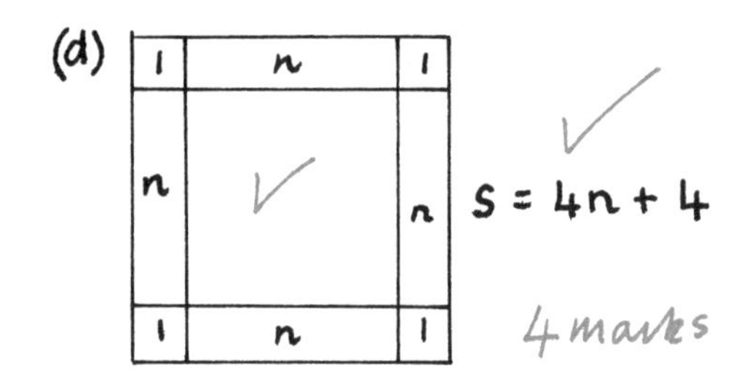

If you multiply the size of the square by four to find the perimeter, then add 4 slabs for the corners. ✓

(e)

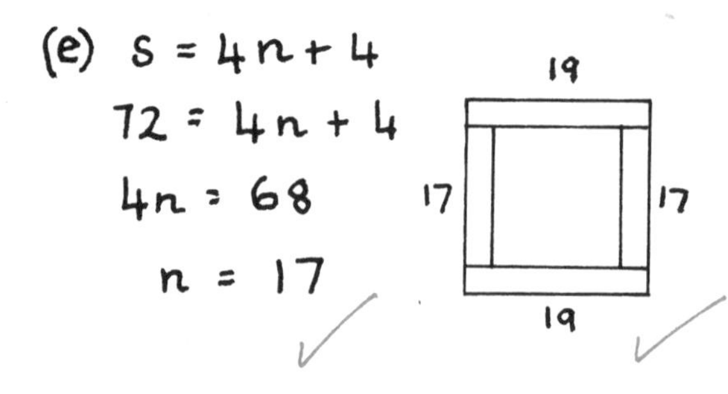

Seventy-two slabs will surround a 17m × 17m lawn. ✓

2 marks

Victoria has answered the problem very well indeed. She gets full marks for a complete table and explanation of her algebraic rule. She has checked her results. Well done!

Final Mark $\frac{12}{12}$

Practising problem solving

It is worthwhile trying to solve a few problems by yourself. Make an attempt at the following. As you search for a starting point, remember the key strategies that we have discussed and keep in mind the *checklist* mentioned on page 10.

Problem 3 Diagonals across rectangles

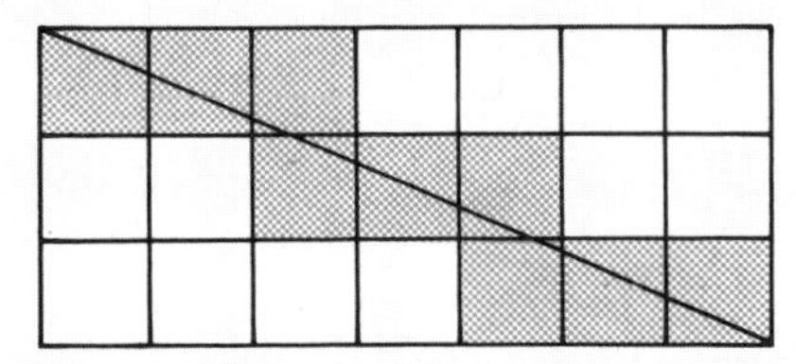

This rectangle was drawn on squared paper and a diagonal line drawn across it.

(a) How many squares did the diagonal line pass through?

(b) Is it possible to predict the number of squares that the line passes through if you are given the dimensions of the rectangle?

Problem 4 Airport control

We shall say that there are 200 major airports around the world, each with a direct flight path *to* all the others, and with a direct flight path *from* all the others.

Imagine that you have a job in Air Traffic Control. Work out how many different direct flight paths there are altogether, connecting all 200 international airports.

Problem 5 Blue cubes

Small unit cubes (1 cm by 1 cm by 1 cm) are joined together to make a larger cube measuring 3 cm × 3 cm × 3 cm. The large cube is then painted blue on all the six outside surfaces. If the cube is then broken down into unit cubes once more, how many cubes have:

6 blue faces?
5 blue faces?
4 blue faces?
3 blue faces?
2 blue faces?
1 blue face?
0 blue faces?

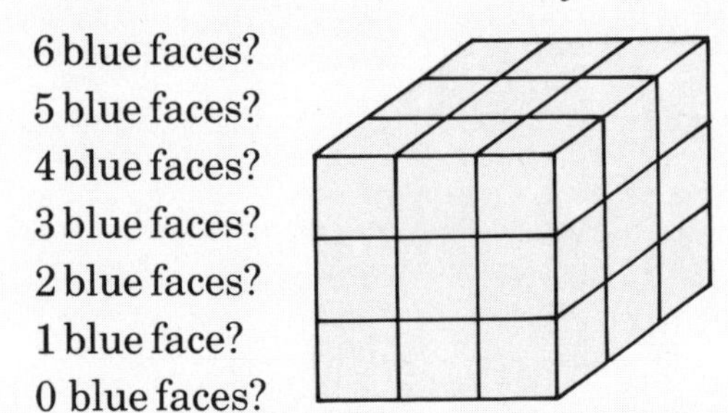

Now repeat the operation with a larger, 4 cm × 4 cm × 4 cm cube.

How many of *each type* of blue faced unit cubes will there be if an even larger 5 cm × 5 cm × 5 cm cube is painted?

Can you predict how many of each type of blue faced unit cube there will be for a painted 8 cm × 8 cm × 8 cm cube?

Now, suppose you have a larger painted n cm × n cm × n cm cube. Can you find a way of forecasting how many of each type of unit cube there will be in this general case? Express the rules symbolically and using your own words.

Problem 6 Matchsticks in squares

Stage 1

Stage 2

Stage 3

Stage 4

The diagram shows a pattern of matches arranged to form squares. Work out how many matches are needed to change from one stage to the next.

Use your answers to predict how many matches will be needed for Stages 5 and 6. How many matches will be used for Stage 20 of the pattern?

Can you predict how many matches will be used for Stage *n* of the pattern?

Problem 7 Timetabling

Problems 7 and 8 do not use the key strategies.

Merry Hall High School employs six teachers of mathematics. The teachers are:

Mrs Ashcroft (Head of Department); Mrs Broad; Mr Hill; Mr Lewis; Mrs Robson; Mr Williams.

Problem 7– Timetabling is best started using trial and error. Then you can reach your final solution by a process of elimination.

The school timetable for mathematics, which is printed overpage, is given to Mrs Ashcroft who has to allocate teachers to all of the classes. *Each class* can be taught by *one teacher only*.

The school timetable has 25 one-hour periods during the week and each teacher must be allowed at least three free periods per week to do his or her lesson preparation and marking.

Show how Mrs Ashcroft can allocate classes, in a fair way, to the individual teachers, including herself.

Day	Period						
	1	2		3		4	5
Monday	6L 3A 3B 3C	1A 1B 1C 3D 3E 3F	*B*	2A 2B 2C 4D 4E 4F	*L*	1D 1E 1F 1G 5A 5B	2D 2E 2F 5C 5D
Tuesday	1A 1B 1C 3D 3E 3F	4A 4B 4C	*R*	2D 2E 2F 5C 5D	*U*	2A 2B 2C 4D 4E 4F	1D 1E 1F 1G 5A 5B
Wednesday	6L 3A 3B 3C	2A 2B 2C 4D 4E 4F	*E*	4A 4B 4C 5A 5B	*N*	2D 2E 2F 5C 5D	6L
Thursday	4A 4B 4C	1A 1B 1C 3D 3E 3F	*A*	6L 3A 3B 3C	*C*	1D 1E 1F 1G 5A 5B	2D 2E 2F 5C 5D
Friday	3A 3B 3C 6L 5C 5D	4A 4B 4C	*K*	2A 2B 2C 4D 4E 4F	*H*	1D 1E 1F 1G 5A 5B	1A 1B 1C 3D 3E 3F

Problem 8 can be solved using graphical representation.

Problem 8 The Easter Egg machine (higher level)

Chicko Eggs Ltd have a machine that makes their Easter Eggs. When it is working at maximum efficiency the machine turns out 1000 Easter Eggs per hour. The machine slows down linearly in production until, after 20 hours, it grinds to a halt. It takes the foreman one hour to service the machine before it returns to maximum efficiency.

The diagram below represents the output of the machine over a 50-hour period.

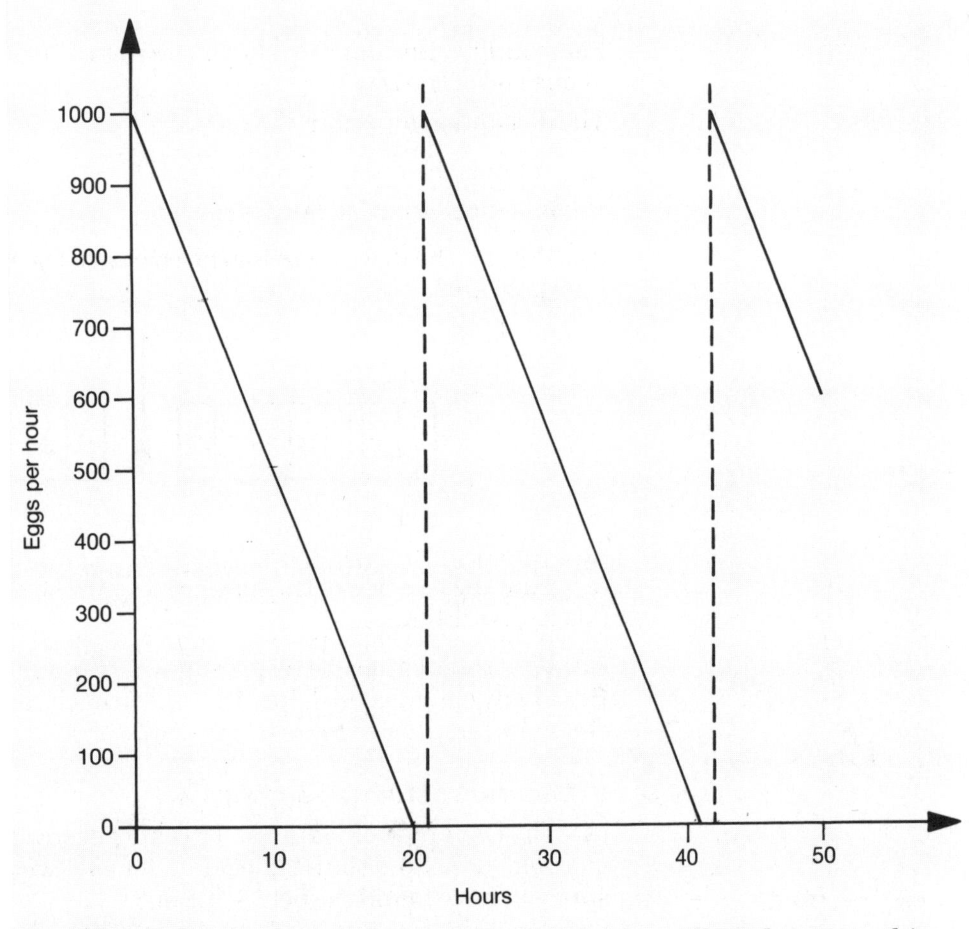

(a) Can you work out how many eggs the machine will produce over this 50-hour period?

(b) This is not the most efficient output for the machine. Draw a diagram to show the machine's output if the foreman services it after 10 hours of production (the service still takes one hour to perform). The effect of this is to keep the output higher, but more time is lost for servicing. How many eggs will the machine produce over a 50-hour period?

(c) Now draw a diagram to show the output of the machine if the foreman services it every five hours of production.

Can you suggest, approximately, how often Chicko Ltd should service the machine to have the greatest production over a 50-hour period?

Problem 9 Passageways

An investigation into the location of passageways in cinemas, etc.

1 Location of one passageway

The Astoria cinema has 10 seats in each row and room for only one passageway. Where in the row should this passageway be located for minimum disruption?

First we need a way of measuring **disruption**. To measure disruption, we shall assume that each person leaves his or her seat once and the disruption is the number of seats he or she has to pass in order to reach the nearest passageway. The disruption for a row is the sum of the disruptions for each seat in that row.

For example, the diagram shows the row of 10 seats with the passageway at one end of the row.

The disruption for seat E is 4.

(a) What is the disruption for seat H?

What is the disruption for the row?

(b) What is the disruption for the row with the passageway located as in this diagram?

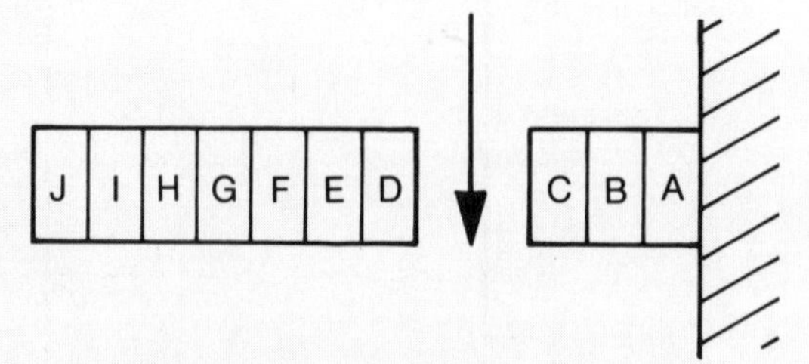

(c) Find the location of the passageway in the above row for minimum (least) disruption.

(d) Find the location of the passageway for minimum disruption in the Broadway cinema which has 12 seats in each row and in the Capital cinema which has 15 seats in each row.

2 Location of two passageways

(a) The Dominion cinema has 12 seats per row and room for *two* passageways. Find the best location for the two passageways. (Are you sure you have found the best solution?)

(b) The Empire cinema has 14 seats per row and room for two passageways. Find the best location for the passageways.

(c) The Flicks cinema has 15 seats per row. Where should its two passageways be located?

(d) And for the Gaumont which has 16 seats per row? And how about some others?

3 Location of three passageways

(a) The Hollywood cinema has 12 seats per row and room for three passageways. Where is their best location?

(b) The Imperial (13 seats per row), the Jupiter (14 seats per row), the Kings (15 seats per row) and the La Scala (32 seats per row) also have room for three passageways. Where should they be located in each cinema? You may care to investigate some more with your own choice of number of seats per row.

What is an investigation?

Investigations are not too dissimilar to the type of problems that you have already attempted in this Section. Whereas in problem solving there appears to be the 'right answer' which you attempt to reach through various methods and using different strategies, an Investigation can involve all of these things *plus* the exploration of other avenues.

To illustrate the *difference* between 'problems', 'investigations' and 'investigative work', an example of each is given below.

Problem

Take a single square. This is stage 1.

Add squares to all outside edges to make stage 2.

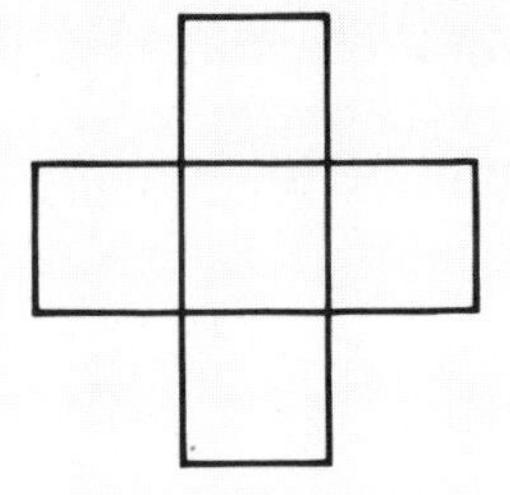

To this shape add more squares to all outside edges to make stage 3.

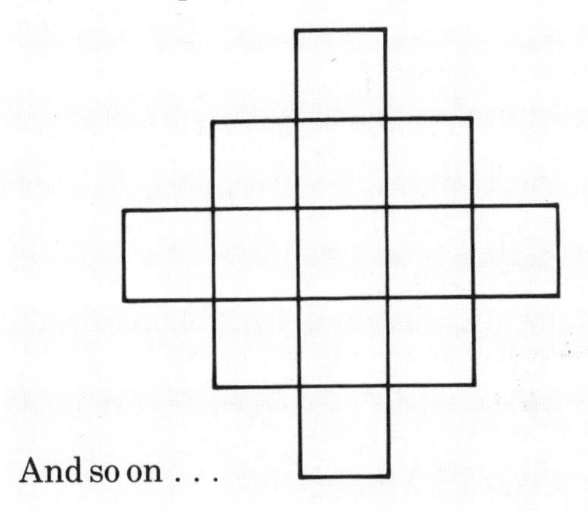

And so on . . .

(a) How many squares are there in stage 15?
(b) How many squares are there in the nth stage?
(c) What stage uses 145 squares?

As you can see, this **problem** is looking for three definite answers. The following **investigation**, however, gives you the chance to study other shapes and does not ask specific questions.

Investigation

Take a single equilateral triangle. Add other equilateral triangles to all the outside edges to make stage 2. Add more equilateral triangles to all the outside edges in stage 2 to make stage 3. And so on . . .

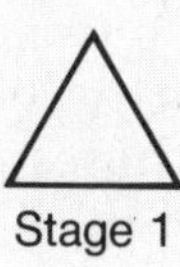

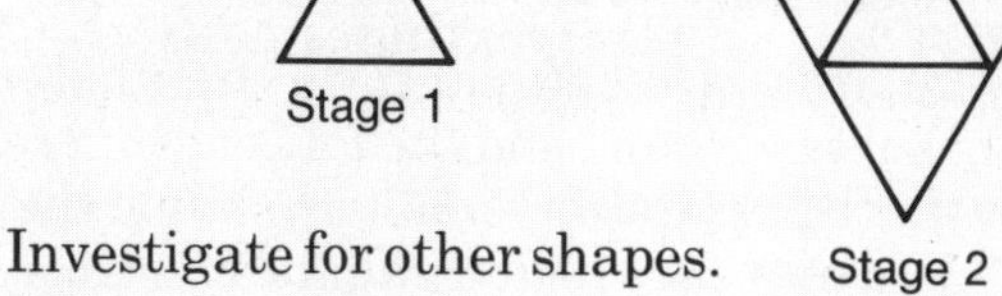

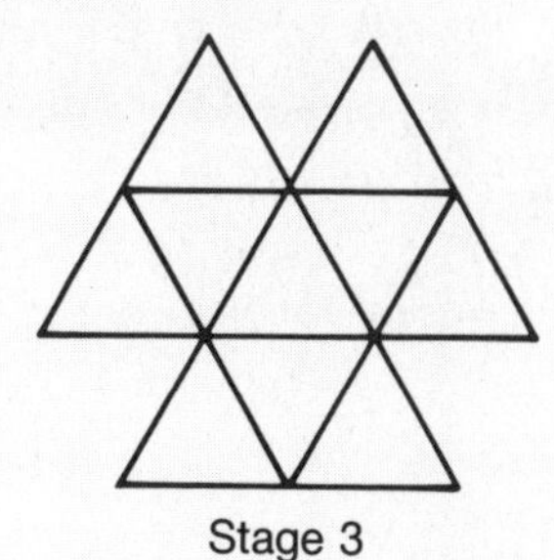

Investigate for other shapes.

Investigative work

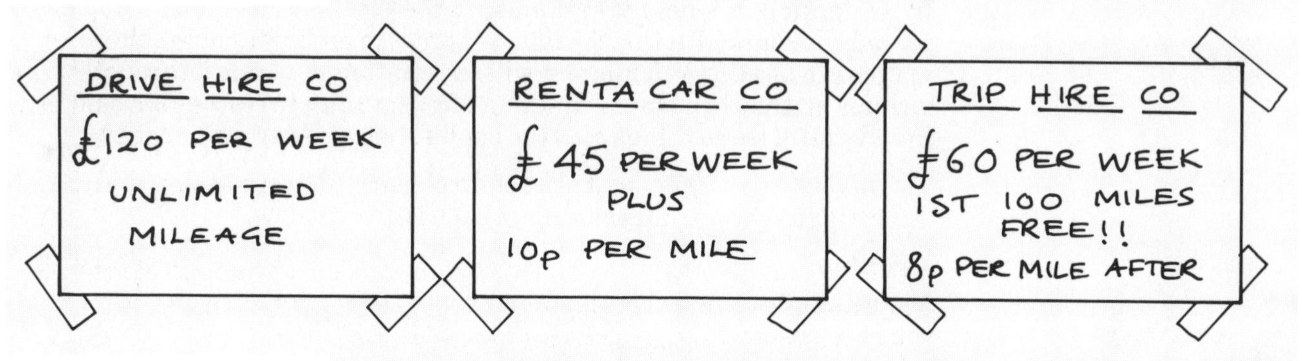

Investigate the hire charges for these car rental firms.

Investigative work usually requires some research.

An investigation presents an open-ended situation. For the student there is a starting point but no finishing line. Students are not expected to produce the correct answer but are required to explore possibilities, to develop mathematical ideas along the way and to be able to explain to others what they find. The emphasis is on exploring the topic in any direction and it is the journey, not the destination that will get you your marks.

What makes a good investigation?

- It must have a reasonably simple starting point.
- It has to contain points where you can develop the investigation along open-ended avenues.
- It must encourage you to *think* in a mathematical way.

I do not intend to narrow your thinking in investigations by giving you rules to follow, as the idea is to be free to explore different avenues in an open-ended way.

‘If you are not sure what type of assignment you are doing, then check with your teacher before you go any further.’

Obtaining solutions is not the aim of investigations, rather the way you develop your line of investigation. Nevertheless, your teacher will be looking for certain aspects of your work in order to award you marks, so you should bear in mind, when tackling any investigation:

- understanding the problem
- planning
- content and development
- communication.

Understanding the problem

At the beginning of your investigation you should write a *clear* statement of the problem. Then you should identify the most important features to be contained in the investigation. Finally, you should state where extensions to the problem might be explored, mentioning any related problems that could arise.

‘Keep your key strategies in mind, although you could start your investigation not quite knowing where you will end.’

Planning

Firstly, you should identify the areas of investigation and state what questions need to be answered.

Secondly, you should research the necessary information, data or statistics required to answer those questions. Check with your teacher to see if the maths department or the library have any books or magazines to help you.

Finally, if possible, you should try to identify alternative strategies for the purpose of comparing results.

Content and development

The best investigations will demonstrate your ability to choose the most appropriate methods to represent your information, and to make relevant and varied calculations leading to the drawing of sound conclusions. Your content must be accurate, appropriate to the problem and contain references and diagrams to support your conclusions.

Communication

Here you must produce a clearly stated conclusion with suggestions for extensions to the investigation. You must also state clearly to what extent your results are valid or incomplete and then go on to discuss the avenues of further investigations.

In making all your statements you should try to demonstrate your own command of mathematical language and concepts and you should present your work in a logical, clear and concise fashion.

‘Asking yourself, ‘what would happen if . . . ?’ could lead you into further investigations or extensions.’

Oral assessment

You may be awarded further marks for your verbal explanations and responses to questioning *both during and after* the completion of the investigation. Your teacher will question you on the areas of the four main aspects we have just discussed. (Oral assessment is dealt with more fully in Section Six.)

Breaking down an investigation

Here is your *checklist*, which you should refer to as you work through your investigation.

- Check vocabulary and relevant information.
- Ask yourself a series of questions.
- Check strategies available and try one.
- Draw diagrams where appropriate.
- Continue with the strategy or change it.
- Write down any results.
- Accuracy: check all workings.
- Always look for patterns in numbers, from tables where appropriate.
- Make predictions.
- Confirm predictions.
- Try to find a general expression, both in your own words and using mathematical symbols where appropriate.
- Communicate your ideas verbally and in writing.
- Check for omissions.
- Look for other avenues by asking yourself questions that begin ‘What happens if . . .’.

You may get an idea of what your teacher is looking for from the comments made.

The following is an outline of the work on an investigation 'in progress'. Some of the points that the student considered are mentioned but not everything is written up.

Investigation 1 Rebounds

Make a rectangle of dots 1 cm apart, measuring 4 cm × 5 cm and label the corners A, B, C and D. Starting at corner A, draw a diagonal line across the squares until it reaches one side of the rectangle. At this point of contact the line rebounds at right angles and continues to another side of the rectangle. Continue rebounding from the sides until you reach a corner.

How many times will it rebound before reaching a corner?

Investigate for various rectangles.

Investigate for different rebounds.

Understanding the problem

Student's thoughts	Student's actions
'I understand the problem. A right angle is 90° and I must remember to include squares in my definition of rectangles.'	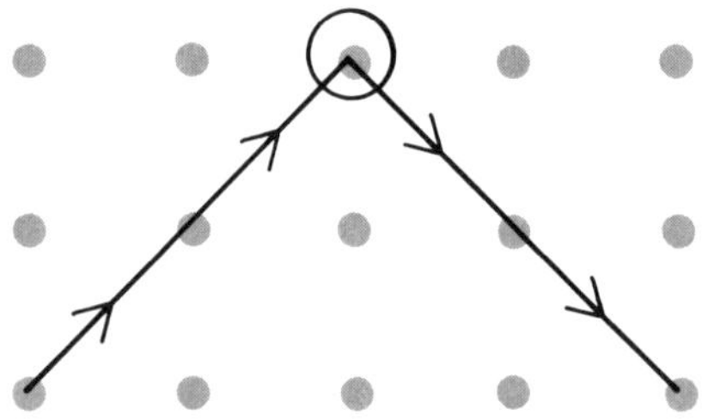 This is a rebound

Planning

Student's thoughts	Student's actions
'Draw the rebounds on a 4 cm × 5 cm rectangle. Then do likewise on a 3 cm × 4 cm rectangle. Then a 2 cm × 3 cm rectangle.'	Collect equipment: **1** Pencil **2** Ruler **3** Rubber **4** Set square **5** Squared paper **6** Checklist

Content and development

Student's thoughts	Student's actions
'I am going to spend half an hour drawing other rectangles and writing down the results for each one.'	A, B, C, D Rectangle 2 cm × 5 cm = 5 rebounds
'I need to be systematic in my drawings. I need a table to show my results effectively. Only then will I be able to spot patterns developing. I am going to draw rectangles of width 2 cm but of different lengths.'	A, B, C, D Rectangle 3 cm × 5 cm = 6 rebounds A, B, C, D Rectangle 2 cm × 3 cm = 3 rebounds
'This investigation is clearly a study of patterns in number.'	
'Are patterns emerging? Can I predict the number of rebounds on a 2 cm × 23 cm rectangle, or a 10 cm × 23 cm rectangle, or a 10 cm × 15 cm rectangle? I can see that squares have no rebounds.'	

Width (cm)	Length (cm)	Number of rebounds
2	2	0
2	3	3
2	4	1
2	5	5
2	6	3
2	7	7
2	8	4
2	9	9
2	10	5

Student's thoughts	**Student's actions**
'The table appears to show that in some cases I can divide the length by the width to find the number of rebounds. If the width does not divide into the length exactly then the larger of the two (i.e. the length or the width) is the same as the number of rebounds.'	
'Wait a minute!!!' 'What about: width 2 cm length 4 cm. According to me the number of rebounds should be 2. Not true!!'	A D B C Rectangle 2 cm x 4 cm = 1 rebound
'What about: width 4 cm length 6 cm. According to me the number of rebounds should be 6. Not true!!' 'RETHINK!!!'	A D B C Rectangle 4 cm x 6 cm = 3 rebounds

Student's thoughts	Student's actions
'If I am going to start predicting the number of rebounds for any rectangle, I had better confirm my predictions. Does a 4 cm × 5 cm rectangle have 7 rebounds?'	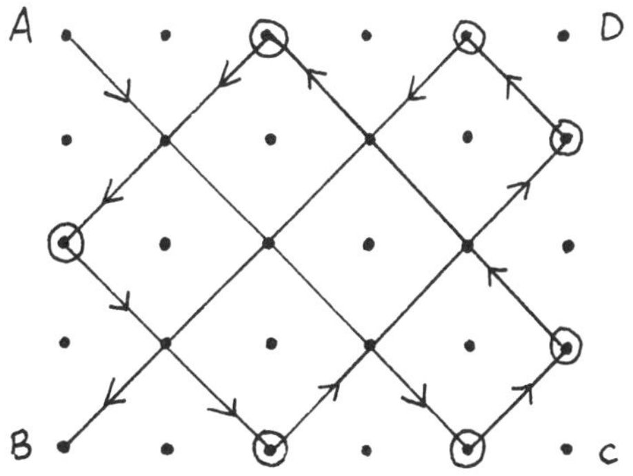 A D B C Rectangle 4 cm × 5 cm = 7 rebounds "Prediction confirmed !!!" Make more tests!
'Can I explain the rules about the patterns in my own words? Can I use algebra to represent my predictions?'	"If I add the width to the length and then subtract 2 I get the number of rebounds. UNLESS :- The width and the length have a common factor. In this case I have to remove this common factor and then apply the general rule to what is left. e.g. 8 × 10 rectangle has $2(4 \times 5) = (4 + 5 - 2)$ rebounds $= 7$ rebounds.
'Can I explain my results to a friend? Can I prove my results? Have I missed anything out?'	Confirm this by drawing!"

Student's thoughts	Student's actions
'I must start thinking about extensions to the investigation.'	A, B, C, D (diagram)
'I am now going to investigate the length of the line in travelling from corner A to the destination corner.'	Destination C. Length 15 diagonals "Look back at previous drawings"
'Can I predict what corner I am going to arrive at if I start from corner A?'	(table below)
'What happens if I use triangular paper and rebound through an angle of 60°?'	"Already I can see a relationship between the length of the line and the length and width of the rectangle."
'What happens if I rebound through a different angle?'	"I need to extend the table to make conclusions about the destination corner.
'What happens if I try the investigation in three dimensions?'	None appears to arrive at D."

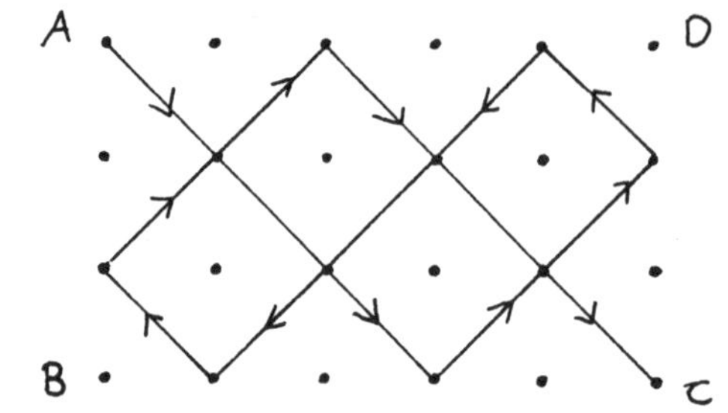

Length of rectangle	Width of rectangle	Length of line	Destination Corner
2	3	6	B
2	5	10	B
2	6	6	C
2	7	14	B
3	7	21	C

Communication

If it is allowed, talk about your investigation to your teacher and your friends.

You should show all your workings – whether right or wrong, as you can thus show that you have tested your predictions and, if you have predicted wrongly, proved them to be incorrect.

Make good use of tables, graphs, models etc to explain your reasoning.

Try explaining your general rule using your own words. Then try to write it down using algebra – after all, algebra is only a form of shorthand for the written word. As an example:

'Square each number and then take away the same number from the result', is the same as:

$$n^2 - n$$

And this is the same as:

'Multiply one number by one less than that number', which is the same as:

$$n(n-1)$$

Finally, you must look back over your investigation to tie up any loose ends.

- Perhaps to confirm a prediction by extending the table.
- Perhaps to prove a prediction wrong by giving an example.
- Perhaps to try another extension to the investigation.

A student's investigation

Now lets follow another outline of the work 'in progress', by a student on a different investigation.

Note This is not the entire investigation. I have taken out extracts, with the teacher's comments, to highlight the important features.

Investigation 2 Joining regular polygons

Ten squares, each of side length 1 cm, are joined together, edge to edge, to form different shapes. Two such shapes are shown below.

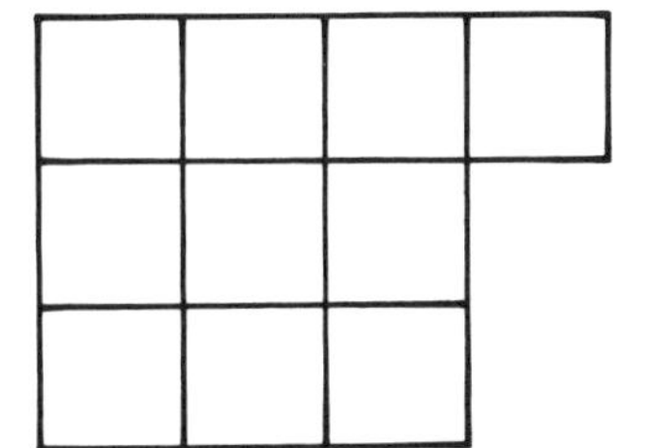

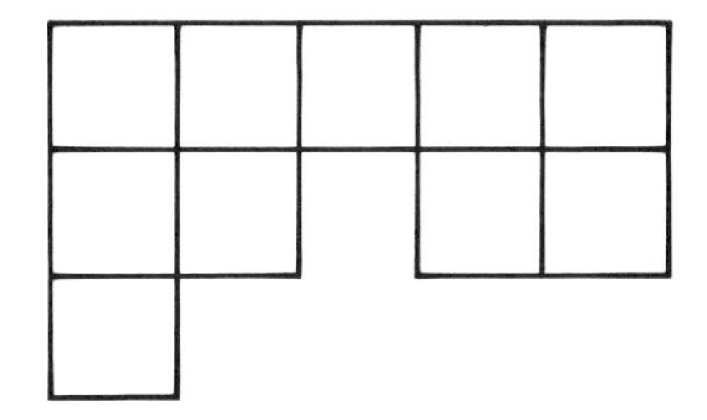

The point at which four squares meet is called a **node**.

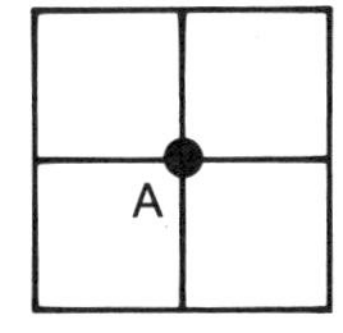

Point A is a **node**

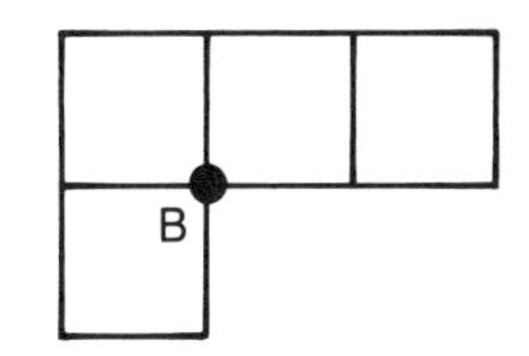

Point B is *not* a **node**

1 For each shape, investigate the relationship beween the perimeter and the number of nodes included in that shape.

2 Making shapes from other than squares, investigate the relationship between the perimeter of each shape and the number of nodes.

Teacher's comments Name Joan Brown

Understanding the problem

This candidate clearly understands the investigation

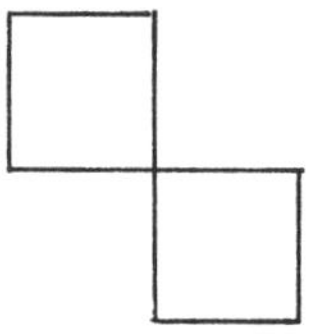

This is not an edge-to-edge join.

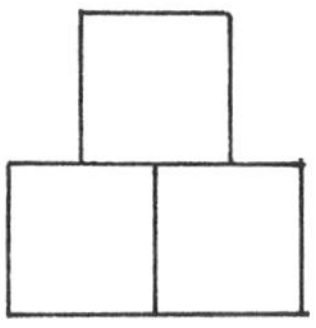

Overlapping is not allowed.

She has set out to explain everything and has not assumed that the teacher understands the task in hand.

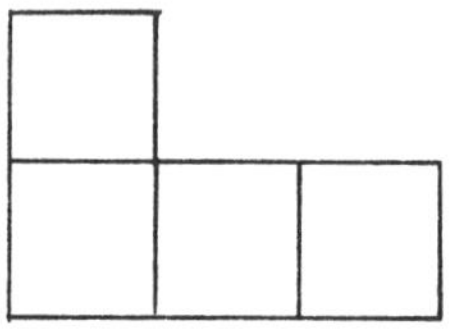

This is a satisfactory edge-to-edge join.

Square

Good choice of regular polygons for tessellating shapes – but do they have to tessellate?

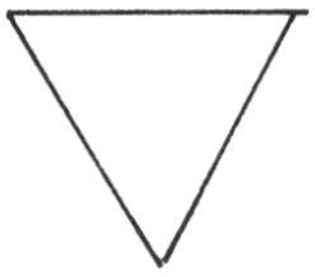

Equilateral triangle

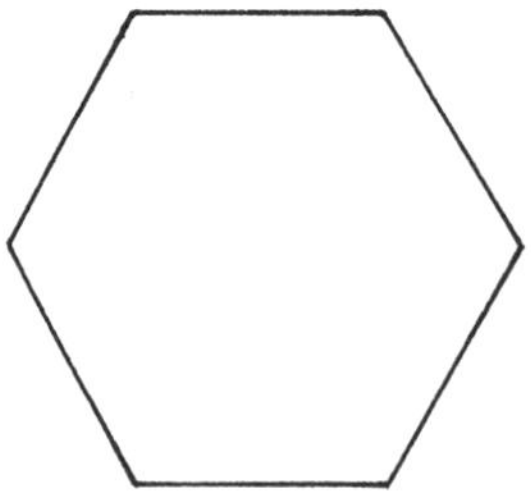

Hexagon.

Explanation of shapes having to tessellate to form a node could be improved with the aid of a diagram.

These are the three regular polygons that will tessellate to form a node that I will use in the second part of the investigation.
The perimeter of any shape is the total distance around the outside. I shall use 2 × 1 rectangle as a non-regular polygon.

Teacher's comments

Summary

5 cm

2 cm

Perimeter = 14 cm
Nodes = 4

Good choice of strategy.
(i) start with a simple case.

I am moving one square to begin with.

(ii) be systematic

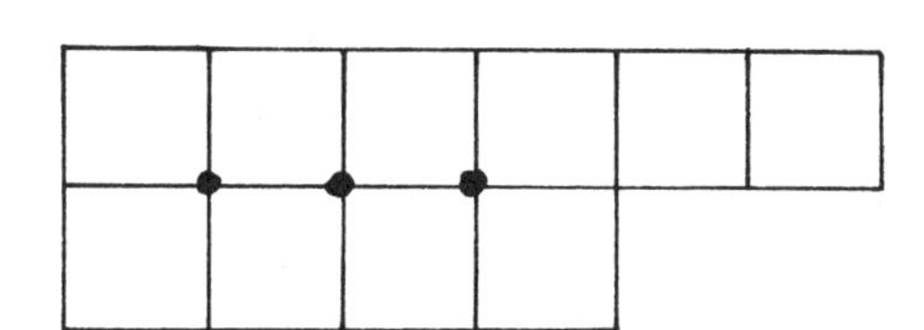

Perimeter = 16 cm
Nodes = 3

If I move the same square to another place.

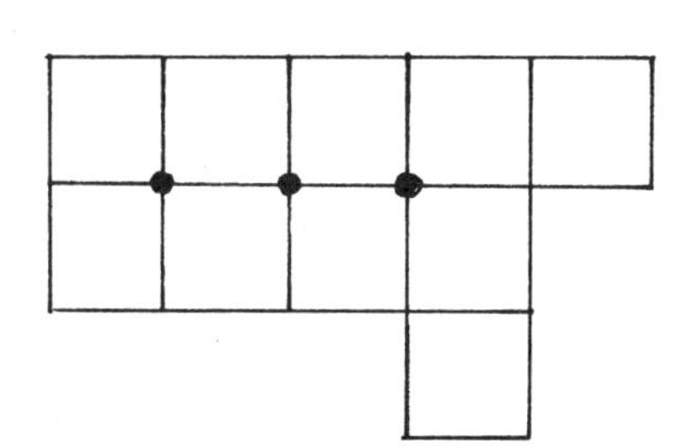

Perimeter = 16 cm
Nodes = 3

(iii) make a prediction

Prediction

No matter where you move this square the results will stay the same.

Prediction should have been tested here.

Now move two squares.

Good systematic approach to the problem.

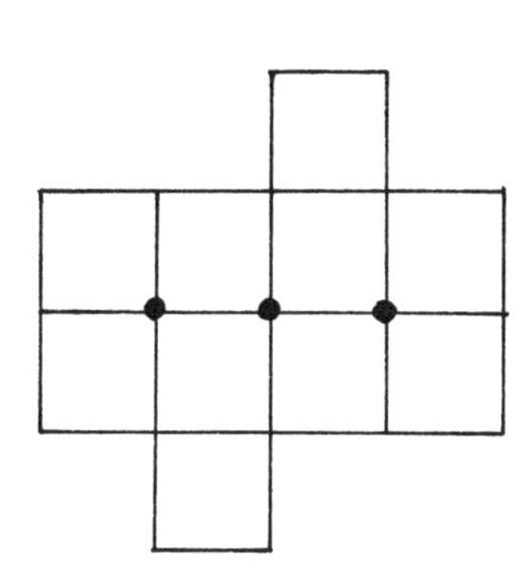

Perimeter = 16 cm
Nodes = 3

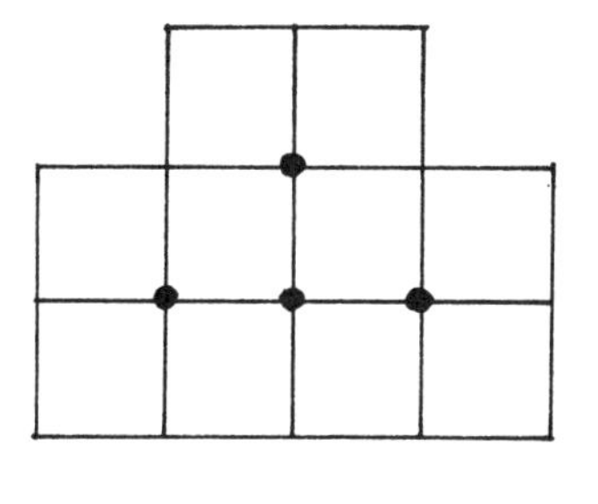

Perimeter = 14 cm
Nodes = 4

Teacher's comments

Joan should have explained why she calls this her final shape.

My final shape will be :-

10 cm

1 cm

Perimeter = 22 cm

Nodes = 0

Good strategy again.

Table of results

Perimeter (cm)	14	16	18	20	22
No. of nodes	4	3	2	1	0

I can see a pattern to my table.

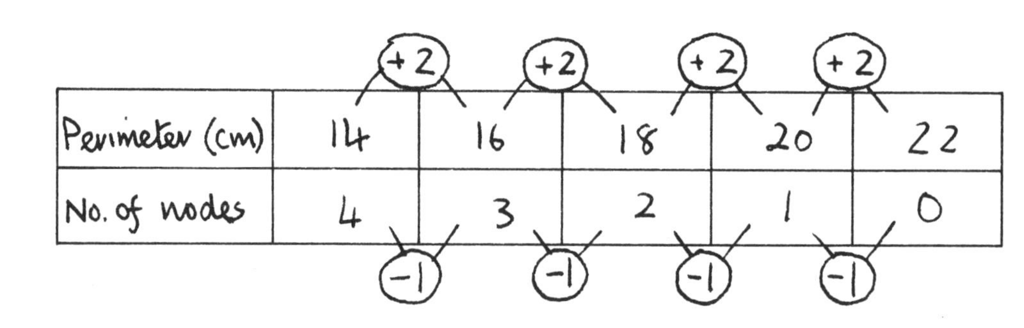

Perimeter (cm)	14	16	18	20	22
No. of nodes	4	3	2	1	0

And again

Generalizing both in words and using algebra shows a good understanding. Solution is very clear.

Finding a rule

The perimeter added to twice the number of nodes always came to 22.

$p = 22 - 2n$ or $p + 2n = 22$

where p = perimeter and n = number of nodes.

Why?

This is not the only case where the perimeter is 14 cm.

Other Comments

Maximum perimeter = 22 cm

minimum perimeter = 14 cm (when squares are arranged in a rectangle).

Teacher's comments

Checking the rule is essential.

Testing the rule

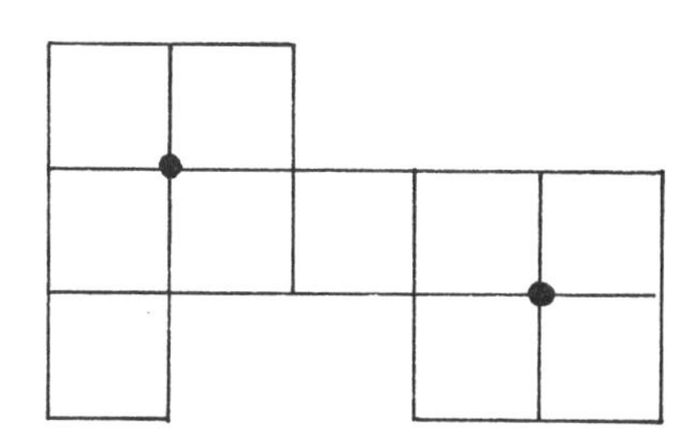

This shape has a perimeter of 18 cm so I predict 2 nodes.
The rule works.

I should have liked to have seen more comments about the shapes formed. How many different shapes were there? Joan should have summarized part 1 of the investigation before going on to 'What happens if...?' section.

What happens if.....?

If I use ten equilateral triangles
Perimeter = 8 cm
Nodes = 2

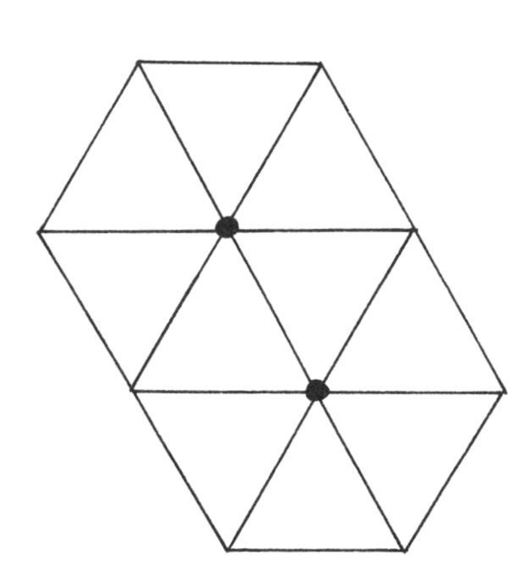

Perimeter = 10 cm
Nodes = 1

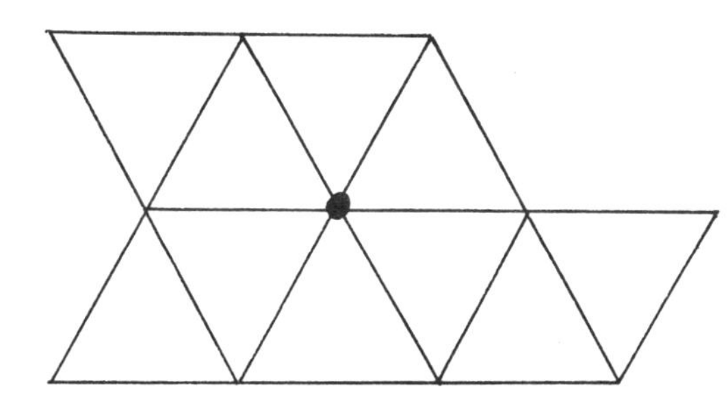

Again, a good choice of strategies.

Perimeter = 12 cm
Nodes = 0

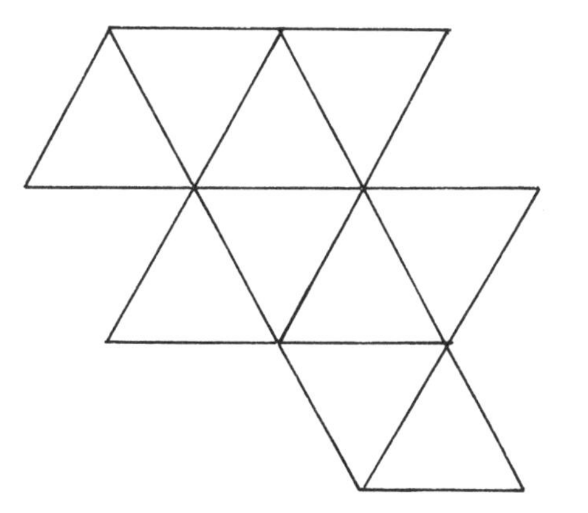

More diagrams needed before a table can be drawn up.

Perimeter (cm)	8	10	12
No. of nodes	2	1	0

Rule: $2n + P = 12$

Joan could have made a comparison with the investigation in part 1.

Teacher's comments

Good extension and good approach

Neat diagrams and easy-to-follow layout.

Very good combination of the previous investigation and choice of strategies.

Prediction confirmed.

What happens if I use hexagons?

Perimeter = 42 cm Nodes = 0

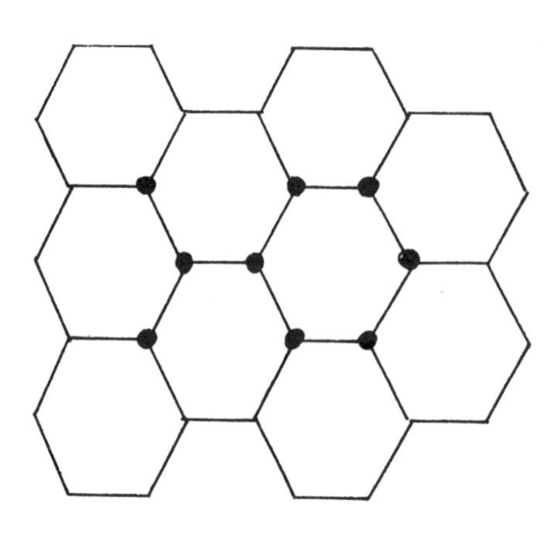

Perimeter = 24 cm
Nodes = 9

Perimeter = 22 cm
Nodes = 10

I have shown the maximum and minimum perimeters
Using the previous investigation, I predict the following table.

Perimeter (cm)	22	24	26	28	30	32	34	36	38	40	42
No of nodes	10	9	8	7	6	5	4	3	2	1	0

Double the number of nodes added to the perimeter always equals 42.
OR $2n + p = 42$.

Test case

Perimeter = 32 cm
Nodes = 5 (prediction confirmed)

What happens if I use a 2 x 1 rectangle?

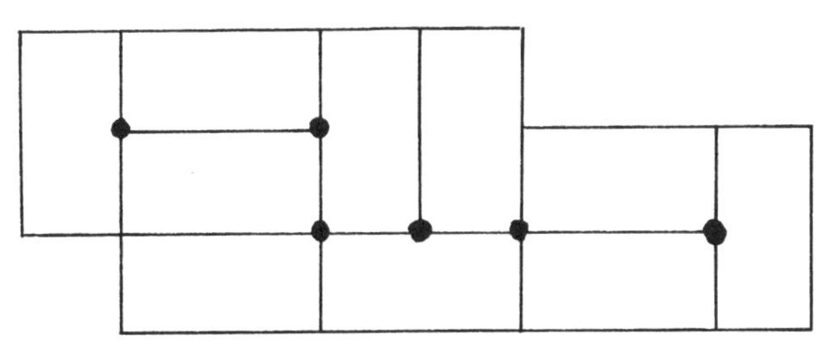

Perimeter = 22 cm Nodes = 6

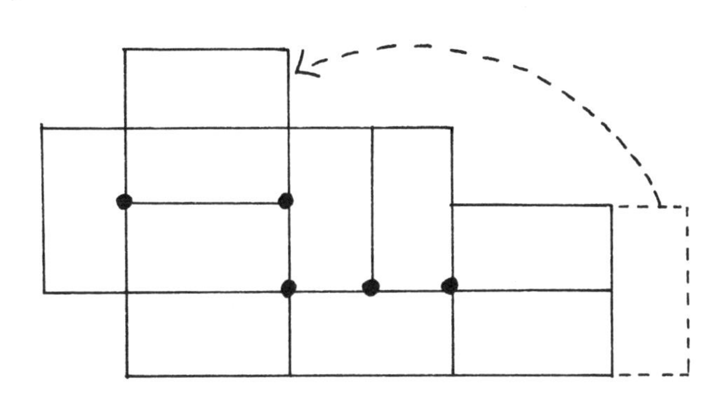

If I move the rectangle on the right of the diagram to another position then:

Perimeter = 22 cm Nodes = 5

This result is due to the fact that there are two different types of node.

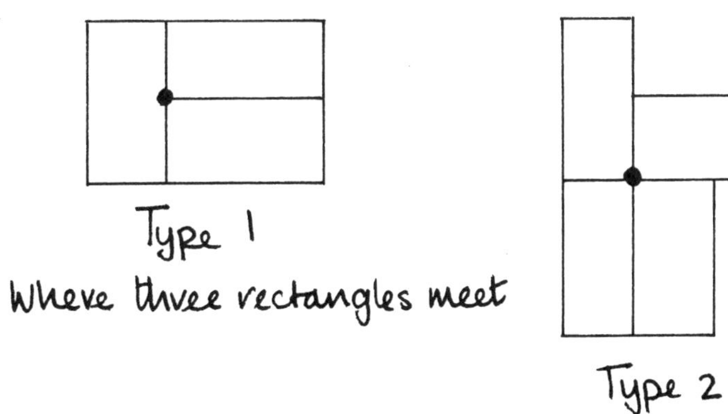

Type 1
Where three rectangles meet

Type 2
Where four rectangles meet

This could be a further extension to the investigation but time will not allow me to follow it.

Teacher's comments

This part is very interesting and should have been followed through.

Joan has done well to spot this change from the patterns developed earlier in the investigation.

Good observation. It is a great pity that this idea was not followed through.

Joan could have improved her summing up of results but nevertheless would have scored highly on this investigation.

Overall presentation is very clear.

Ideas for investigations

Here are some ideas for investigations that you might like to think about. If you cannot afford the time to complete them, you might like to produce your own summaries. Suggested summaries appear in Section Seven, but remember:

- investigations are meant to be open-ended
- you should explore your own avenues of further investigation and not be influenced totally by the summaries
- if you feel that you have done enough, discuss the situation with your teacher to see if you are in a position to stop the investigation.

Investigation 3 Consecutive sums

$$13 = 6 + 7$$

The number 13 can be written as the sum of two consecutive numbers.

$$15 = 7 + 8 \quad or \quad 15 = 4 + 5 + 6 \quad or \quad 15 = 1 + 2 + 3 + 4 + 5$$

The number 15 can be written as the sum of two consecutive numbers, three consecutive numbers or five consecutive numbers.

$$10 = 1 + 2 + 3 + 4$$

The number 10 can be written as the sum of four consecutive numbers.

Consider all the numbers from 1 to 30.

(a) Are there any numbers that cannot be written as a sum of consecutive numbers?

(b) Which numbers can be written as a sum of consecutive numbers in more than one way?

(c) Can you *predict* whether any number can be split into consecutive sums?

(d) Can you predict in *what ways* a number can be split into consecutive sums?

(e) Consecutive might not always mean successive numbers.
What about consecutive odd numbers?
What about consecutive even numbers?
Investigate.

Investigation 4 Surrounded by the 'blue tide'

On 1 cm squared paper draw a 3 cm × 3 cm square. Colour the small squares just inside the perimeter blue, and the middle one black.

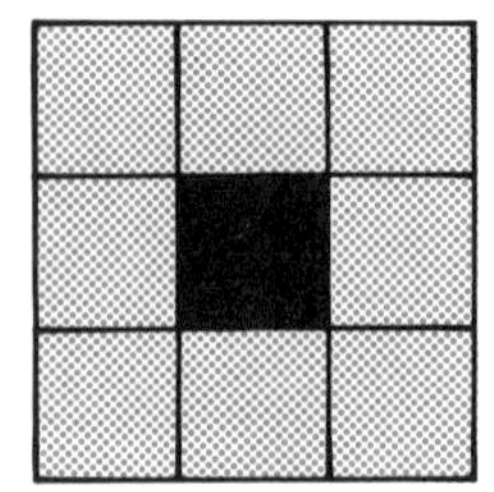

Repeat for a 4 cm × 4 cm square, and so on . . .

Are there relationships between the area, the perimeter, the number of blue squares and the number of black squares?

Investigate for other shapes.

Investigation 5 Crossing lines and regions

When *two* straight lines cross, they have *one* crossing point and form *four* regions.

When *three* straight lines cross, they have *three* crossing points and form *seven* regions, *one* of which is *closed*.

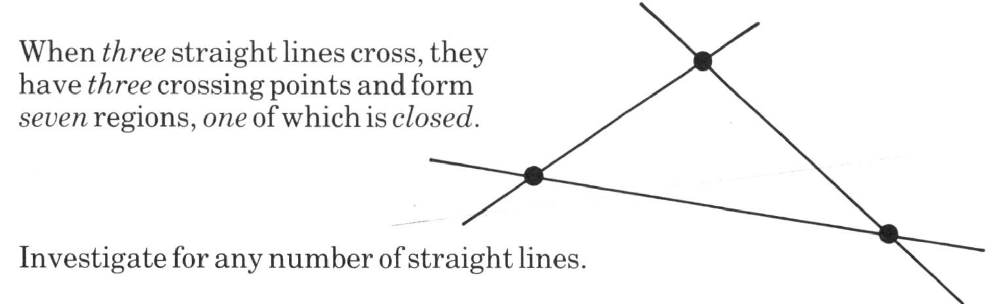

Investigate for any number of straight lines.

Investigation 6 Changing places – 'Frogs'

Blue and black counters are placed at either end of a line of squares.

Blue counters can only move to the right by sliding or jumping.

Black counters can only move to the left by sliding or jumping.

Counters may jump over counters of different colour, one at a time, only providing there is a vacant space to land upon.

These two counters can change places in three moves.

Move 1 Black slides left

Move 2 Blue jumps over black

Move 3 Black slides left

Show how the two blue and two black counters can change places in eight moves (4 jumps and 4 slides).

Investigate for any number of counters.

What happens if the number of blue counters is not the same as the number of black counters?

Investigation 7 Palindromic numbers

A **palindromic number** is one which reads the same, backwards and forwards. For example, these are all palindromic numbers:

44
151
818
3553

Now, choose any number, say, 65
reverse the digits, 56
then add them. 121

Is the answer a palindromic number? If not, as below,

37
73
110
reverse the digits in the answer, 011
add the new answer to the old answer. 121

Is the new answer a palindromic number?

Investigate for different numbers less than 100.

Are there palindromic numbers between 100 and 1000?

SECTION THREE

Statistics

What is 'statistics'?

Before you start your statistical assignment you should familiarize yourself with what statistics is all about.

Statistics is the name given to the collection of facts and the studying and analysis of them. These facts are referred to as **data**.

In the modern world, statistics has become the study of any kind of data, so that we can use the information to help explain, understand and predict trends.

For example, every 10 years, the Government spends millions of pounds collecting and analysing the **National Census**, for which every householder in the country is obliged to complete a Census form. This contains a variety of questions about people living in the house at the time of the Census. Institutions such as hotels, jails, hospitals etc also complete the Census form, so that every person in the country at the time of the Census has been accounted for. The Government then has excellent data about the population which it can use in plans for the future.

Statistics are all around us: newspapers, advertisements, sports etc all use facts and figures in some form. Scientists collect data from their experiments to try to help them test new drugs. Businesses and industry use statistics to help them judge how successful they are or whether a business proposition is realistic. Schools continually use statistics to measure your progress.

Carrying out a statistical assignment

If you examine the flow chart you will get some idea as to the basic stages involved in a statistical assignment. The rest of Section Three explains the stages and you will need to refer back to the flow chart, from time to time, to see where each stage fits into the assignment as a whole.

'By stating a hypothesis you have given yourself a target. It should help to guide you through your survey and give you something to prove or disprove.'

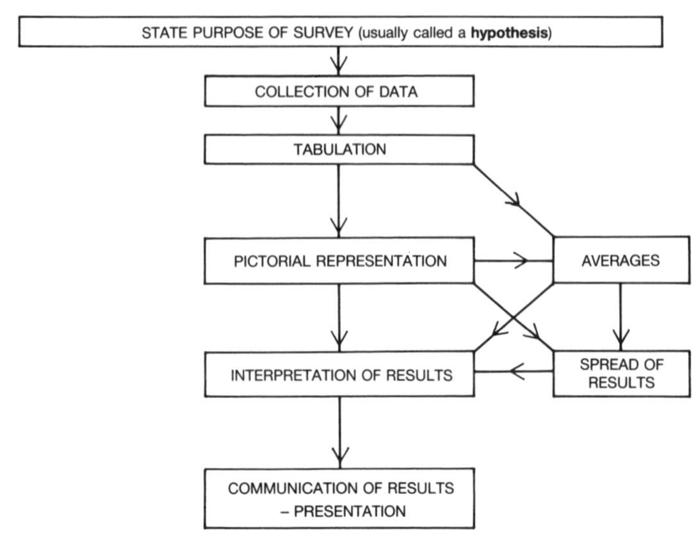

State hypothesis

It is vital that you give a **purpose** to your survey. It is pointless to collect data for no reason.

State what you hope to achieve at the very beginning of your assignment.

Collection of data

Data can be collected in a number of ways. The main ways would be:

- by a census
- by random sampling
- by observation
- by experiment
- from published information.

Census

As already mentioned, the National Census is a very expensive undertaking and takes a long time to process.

You could, however, attempt your own census, restricting the number of people taking part to a manageable size e.g. all the members of your class, form, year, street, youth club etc. Make a list of questions that will result in simple numerical answers, such as:

(a) How old are you?

(b) How many people are there in your family?

(c) How much pocket money do you get each week?

(d) What wages do you get paid from your part-time job per week?

(e) How many hours do you work per week?

(f) How many hours homework do you do each week?

(g) How many pets do you have at home?

Asking each person individually would take time, but would be more likely to produce information you seek. If you send out the questions on forms, many may not be returned to you, completed.

Random sampling

Random sampling is similar to a census but it involves questioning people *at random*, rather than questioning a whole group.

When constructing your **questionnaire**, or list of questions, care must be taken with the questions so that they are clear, precise, require short answers and are not ambiguous. It is useful to keep the following points in mind.

(a) Do not make your questionnaire too long.

(b) Include all the questions required by the survey.

(c) Keep the language simple.

(d) Phrase questions so that they can have only *one* meaning (i.e. they are not ambiguous).

(e) Phrase questions so that they can be answered by a simple YES/NO, a tick in a box, or a number.

(f) Do not ask questions that might embarrass or upset anyone.

(g) Avoid *vague* words like small, large, tall, big etc.

(h) Avoid questions that involve calculations – people can make mistakes!

If you *send* your questionnaires to the people in your survey, it is likely that less than half will return them. And remember, often the people who do return them are the especially interested people or the dissatisfied people, and this could lead to a biased survey.

It is better to *ask* the questions to the people in your survey and to fill in the questionnaire yourself. This way you can be sure of the answers and get immediate results. Opinion polls and various 'market surveys' are conducted this way, with people being stopped in the street and questioned.

Here is a simple questionnaire which could be used to decide whether a swimming pool is needed in your area.

PLEASE TICK UNLESS REQUESTED OTHERWISE

1 Your age:

5-	10-	15-	20-	30-	40-	50-	65 and over

2

Male	Female

Swimmer	Non-swimmer

4 Would you use the pool to:

Learn to swim?	
For fun?	
For exercise?	
For competitive training?	

5 How many hours a week would you use the pool?

0-2	2-4	4-6	6-8	8-10	Over 10

6 What time of day would you use the pool?

Morning	Afternoon	Evening

Observation

Some data cannot be collected by questioning, e.g. the flow of traffic of various types. To collect this kind of data you must go and observe what happens and note down how many times it happens. To do this you will need to complete a **tally chart**. (You will find a clipboard for your paperwork very useful.)

Suppose you are doing a survey on the flow of road traffic. Your tally chart may look like this. (We shall be explaining tally charts, later in this section.)

TRAFFIC FLOW SURVEY

Carried out by: JANET HARPER
Location: NILE ST BURSLEM
From: 1.0pm to 2.0pm
On: MONDAY 14TH AUGUST 1989

Type of vehicle	Tally	Frequency
CAR	~~||||~~ ~~||||~~ ~~||||~~ ~~||||~~ ~~||||~~ ~~||||~~ |||	33
VAN	~~||||~~ ~~||||~~ ~~||||~~ ||||	19
LORRY	~~||||~~ ||	7
BUS		0
MOTOR CYCLE	|||	3
CYCLE	~~||||~~ ~~||||~~ ~~||||~~ ~~||||~~ ~~||||~~		26
OTHER		0

This is a very popular means of collecting data. However, it does require self-discipline on your part as you will probably be working, often alone, on very busy roads. There will always be the temptation to 'cheat'! And, be prepared for all kinds of weather!

Check that you have all the correct equipment and necessary paperwork with you before starting out on your journey to collect data.

Another practical and simple idea for a survey of this kind is to select two similar businesses that are close together and can be observed from the same point, a little distance away. Such businesses could be garages, banks, newsagents etc. Collect data about the number of customers using each and then say why one is more popular than the other.

When carrying out these types of survey, remember to be considerate to the people being surveyed. A lot of people would not like to think they were being 'spied on'. And also, remember your own safety. If possible, ask a friend to accompany you – just so you are not entirely alone, and make sure a teacher, or parent, or responsible friend, knows where you are and what you are doing.

Experiment

Some data can only be obtained by doing experiments. For example:

(a) For how long can people hold their breath?
(b) How many red 'Smarties' in a tube?
(c) What is the effect of exercise on the pulse rate?
(d) What is the effect of fertilizer on a tray of seeds?

The only way to measure the effect of fertilizer on a tray of seeds is to plant seeds from the same packet, into the same compost, in two separate trays and then treat only one of the trays with fertilizer. As time passes, only then will you be able to observe a measure of differences.

Experimental data can take time to collect and could cost you money. Leave yourself enough time, and be prepared to wait for results.

Published statistics

Many statistics are published regularly in magazines and newspapers and can easily be obtained by those interested. As just one example, American Football statistics can be obtained from weekly, monthly or annual publications. You will, of course, know that the results of our own Saturday's football fixtures fill the sports columns in Sunday's newspapers.

Sampling

If you are *not* collecting your data from the whole of one particular group – called a **population** – then you will have to think about the best way of choosing your **sample**. Beware it is not a *biased* sample, which will give you biased results. Try to aim for a *random* sample which will give you unbiased results.

Suppose you are considering, say, the number of people who use a pedestrian crossing at a particular point in town. *Do not* conduct the count for one hour from say, 12 noon to 1.00 p.m. If you do, your results will give you a distorted picture, as 12 noon to 1.00 p.m. could be the busiest time of the whole day. You will get more realistic results if you make your head-count in 15-minutes spells at four different times of the day.

Similarly, if you are trying to find out by what mode of transport students arrive at school, do not ask the first 50 people who arrive through the school gates. It could be that a bus has just arrived and most of the 50 students got off that one bus. More unbiased results will be obtained by asking every tenth person until everyone has arrived, as this is more of a random process.

If you are taking a sample, then check with your teacher that you are selecting it correctly. You do not want to spend two weeks working with figures that will give you only biased results.

Tabulating data

There are two types of numerical data. The first is called **discrete data**, where you take certain values. 'Counting' is a form of discrete data. The second is called **continuous data**, where you take values within a certain range. 'Measuring time' is one form of continuous data.

Having collected the raw (unprocessed) data that you require, the next thing you have to do is to arrange it into tables to show the results of your survey. The types of tables which we will consider are:

(a) tally charts
(b) frequency distributions
(c) group frequency distributions
(d) cumulative frequency distributions.

Whenever you are transferring raw data into a table, be careful, as it is easy to make mistakes. Neatness and accuracy are important at this stage.

Tally charts

If you are collecting data over a period of time e.g. counting cars, golf scores etc, then you will need to keep a tally. To make a tally chart, make a list down the left-hand side of your page including all the possibilities that you are counting. As each result occurs place a tally stroke (|) in the appropriate space for your result.

It is usual to list tally marks in groups of five. When a row of four tallies have been obtained (||||), then on the fifth, place your tally across the other four (𝍸) to make your group of five, It is easier to count up your total in fives, rather than to count a large number of single tally strokes.

For example: Make a tally chart of the number of people in cars that pass the school gates.

Number of people per car	Tally
1	𝍸 𝍸 𝍸 𝍸 𝍸 |||
2	𝍸 𝍸 𝍸 ||
3	𝍸 𝍸 |
4	𝍸 ||||
5	||
6	|
more than 6	

Frequency distribution

When you have constructed your tally chart, it is usual to put the total of tallies into a separate column on the right-hand side of the tally chart. This then becomes a **frequency distribution**. The frequency of an event is the number of times that an event happens.

Another situation you may be faced with is a page full of figures placed in no order whatsoever. Imagine you have noted down how many brothers and/or sisters each of the 100 first year pupils in your school has. Their answers are the figures listed below.

1 2 0 1 1 2 1 3 4 5 6 1 0 2 1 3 1 2 4 1
2 1 3 1 0 1 2 0 3 1 0 0 2 3 1 2 0 1 2 0
0 0 1 4 2 2 2 1 0 0 1 3 1 2 3 2 1 3 4 5
2 2 1 1 0 1 2 3 1 2 4 3 2 0 0 1 2 3 1 1
1 2 1 0 0 1 2 4 2 3 0 1 2 2 1 1 1 2 0 1

To construct a frequency distribution you must make out your tally chart. As you enter the tally mark, you must cross out the corresponding number on the list. By doing this you will know, at the end, whether you have missed out any. Then you add up all the tallies for each different number of people and place that total in the frequency column.

As a check you should add up the frequency column. This should be the same as the number of figures that were in your list, that is 100 in *this* example.

Brothers and sisters	Tally	Frequency
0	𝍸 𝍸 𝍸 IIII	19
1	𝍸 𝍸 𝍸 𝍸 𝍸 𝍸 IIII	34
2	𝍸 𝍸 𝍸 𝍸 𝍸 I	26
3	𝍸 𝍸 I	11
4	𝍸 II	7
5	II	2
6	I	1
	Total	100

Grouped frequency distribution

The **range** of a set of data is defined as the difference between the highest and lowest values in the data.

The set of scores for Form 1 in a maths test is:

{5 4 7 9 8 3 7 6 5 4 7 8 3 9 6 10 5 4 3 7 }

The **range** = highest score − lowest score
= 10 − 3
= **7**

If you can imagine a set of marks out of 10 for the *whole* school, then we would use a **frequency distribution** to tabulate the marks. However, if the marks were out of 100, then the range would be far greater than 7, as it was in the example. To make a tally chart when the range is so wide would be unmanageable. Instead of having 101 rows in your tally chart to show each of the scores ranging from 0 to 100, it would be more sensible to group the scores together within **class intervals**.

Exam marks out of 100 are usually grouped in the classes:

0-9, 10-19, 20-29, 30-39, etc

The last class interval is 90 and over. Then the table needs only 10 class intervals and will thus use up only 10 lines of your page.

If the marks are grouped together as:

0-19, 20-39, 40-59, 60-79, 80 and over

then the table would need only five class intervals and use up five lines of your page.

Data can also be shown using class intervals of different sizes:

0-19, 20-39, 40-49, 50-59, 60-79, etc

The interval 40-49 has an **upper class limit** of 49 and a **lower class limit** of 40. It also has a **mid-point value** of 44.5. This is arrived at by adding the upper class limit to the lower class limit and dividing the answer by two. In calculations, this mid-point value is used to represent the whole of the class interval 40-49.

When the *totals* for each class interval are added up then we have a **grouped frequency distribution**.

Here are the marks out of 100 for a group of 60 candidates in a general knowledge test:

31 27 58 61 75 33 42 17 57 9 82 66
23 56 75 42 47 61 34 88 71 52 53 61
67 71 47 33 71 90 42 55 17 21 36 47
51 67 75 37 49 70 29 39 51 67 74 49
58 66 62 77 41 38 20 11 57 61 59 81

Using class intervals 0-9, 10-19, etc, the grouped frequency table would look like this.

Class interval	Tally	Frequency
0- 9	/	1
10-19	///	3
20-29	⫽⫽	5
30-39	⫽⫽ ///	8
40-49	⫽⫽ ////	9
50-59	⫽⫽ ⫽⫽ /	11
60-69	⫽⫽ ⫽⫽	10
70-79	⫽⫽ ////	9
80-89	///	3
90 and over	/	1
	Total	60

Using different class intervals the same information can be shown like this.

Class interval	Tally	Frequency
0-19	////	4
20-39	⫽⫽ ⫽⫽ ///	13
40-49	⫽⫽ ////	9
50-59	⫽⫽ ⫽⫽ /	11
60-69	⫽⫽ ⫽⫽	10
70-79	⫽⫽ ////	9
80 and over	////	4
	Total	60

When data is grouped together like this then *complete accuracy* is lost, but if you choose your intervals carefully then the advantages of making your numbers easier to use far outweigh the disadvantages of loss of accuracy.

If you have collected a lot of data with a wide range of variables, then grouped frequency distribution tables are essential. They also give you an opportunity to demonstrate your further knowledge of statistics.

Cumulative frequency distribution

This type of distribution is only included on the higher level syllabus but that need not stop you from trying to use one.

To construct a cumulative frequency table from a simple frequency distribution, you have to add up the frequencies as you move higher up the range that you are considering.

To illustrate this, consider an example of 60 houses being advertised for sale, in the window of a local estate agent.

Cost (in £1000s)	130-	140-	150-	160-	170-179.999
Frequency	16	11	15	10	8

Here, 16 houses are in the price range £130 000-£139 999. A further 11 homes are in the price range £140 000-£149 999, and so on. . . .

In a **cumulative frequency table**, we combine the first two results and say that 27 houses are priced *less than* £150 000. We can go on to say that 42 houses are priced *less than* £160 000, and so on. . . .

Your cumulative frequency table would look like this:

Less than £1000s	140	150	160	170	180
Cumulative frequency	16	27	42	52	60

The final entry shows that all 60 houses are priced at *less than* £180 000. This final figure should always be the same as the total number of frequencies.

Similarly, for the grouped frequency table for class intervals 0-9, 10-19, etc, on page 46, the cumulative frequency table would look like this:

Score less than:	10	20	30	40	50	60	70	80	90	100
Cumulative frequency	1	4	9	17	26	37	47	56	59	60

The reason for using a grouped frequency table is to illustrate how the whole group is progressing rather than one particular set. Also, it helps in calculating averages and in displaying your statistics, using a *cumulative frequency graph*. These will be explained in the next few pages.

If you can successfully incorporate a cumulative frequency distribution in your survey *and understand it fully*, you will be showing a good standard of mathematics.

Pictorial representation

Once you have collected your data and organized it in table form, the next thing that you need to do is present it in an attractive and eye-catching manner. The best way to do this is to put your data into picture form. This is most important, as your diagrams will create an overall impression of the whole assignment. When drawing your diagrams keep the following **checklist** in mind:

(a) be neat and accurate

(b) use colours to make them more attractive

(c) draw as many *appropriate* types of diagram as possible

(d) keep your diagrams clear and simple

(e) make sure all diagrams have a title and are suitably labelled

(f) beware of misleading diagrams.

The basic ways of showing your information would be to use the following:

1 pictograph **2** pie chart **3** bar chart

4 line graph **5** cumulative frequency diagram

You can add your own personal touches to these diagrams and a quick look at any statistical text book will show you variations you can try. We shall now look at the basic ways, in turn.

Pictograph

One of the most eye-catching ways of illustrating statistics is to make your graph in the form of a picture. This kind of graph is called a **pictograph** or **pictogram**.

A firm built a new development of houses during 1988. Their completion rate for the first six months of the year is shown here.

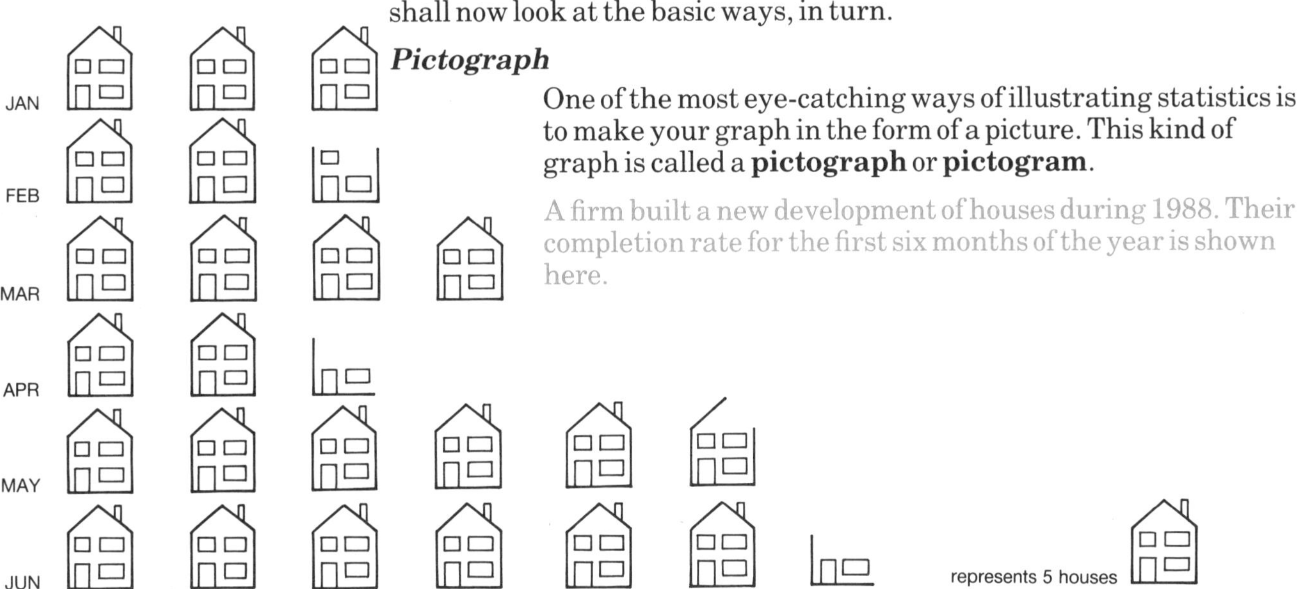

These kinds of diagram have two main drawbacks:

(a) they take a long time to draw well

(b) sometimes, when (as above) they show 'bits' of pictures, they are difficult to read accurately.

Pie chart

Another way of illustrating data is to draw a **pie chart**. This is a circle divided up into sectors which each represent a section of data. The *size* of the sector depends on the *size* of the section of data you are considering.

We shall now work through an example.

30 children from a class were asked how they came to school. The results were:

bus 4; car 2; cycle 6; walked 18

To draw the pie chart illustrating this data, you will need a pair of compasses, a pencil, a protractor, a ruler, coloured pens or pencils and, if possible, a calculator. You then simply follow these rules.

(a) Add up all the frequencies

$$4 + 2 + 6 + 18 = 30$$

(b) Divide 360° (the number of degrees for a whole circle) by the *total* number of frequencies. This gives the number of degrees by which *each item* of data is represented.

$$360° \div 30 = 12°$$

(c) Multiply, in turn, your answer in **(b)** above by the frequency in each **section** of data. This will give the angle of the sector representing each section of data. For example:

$$\text{bus} = 12° \times 4 = 48°; \quad \text{walked} = 12° \times 18 = 216°$$

(d) Add up the angles for *all* the sections of data to check that they come to 360°.

$$48° + 24° + 72° + 216° = 360°$$

(e) Draw a circle of any convenient radius with a pair of compasses.

(f) Draw as a starting line, a radius, usually from the 'twelve o'clock' position.

(g) From this starting line use your protractor to divide the circle up into sectors, using your results from **(c)**.

(h) Label the various sectors: bus, car, etc and shade them using different colours.

For the class of 30, their mode of transport when coming to school was, thus, as follows.

Type of transport	Number of pupils	Angle of sector
bus	4	$\frac{360}{30} \times 4 = 48°$
car	2	$\frac{360}{30} \times 2 = 24°$
cycle	6	$\frac{360}{30} \times 6 = 72°$
walked	18	$\frac{360}{30} \times 18 = 216°$
		Total $= 360°$

The main advantages of using pie charts is that you can easily see one particular section of the data as a fraction of the whole group.

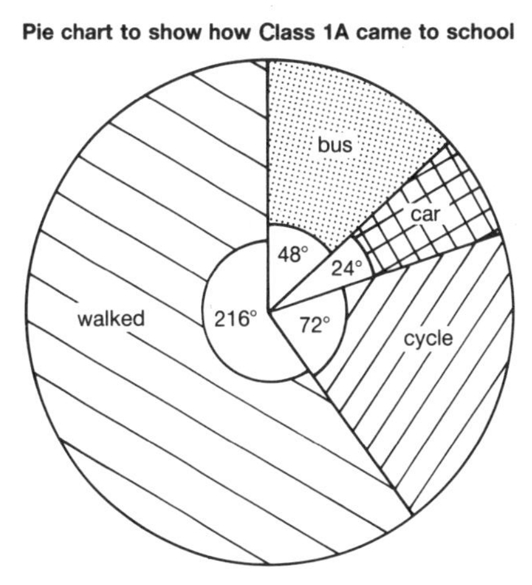

Bar chart

A **bar chart** is made up of a series of bars of *equal* widths, each representing a section of data, in the same way that sectors represented sections of data in a pie chart. In this case, however, the bars are placed side by side and it is the *length* of the bar that represents the size of the section of data you are considering. A *vertical* bar chart is sometimes called a **column graph**.

From the distribution below, draw a bar chart to show the types of drink preferred by a class of 30 boys and 30 girls.

Drink:	Tea	Coffee	Coke	Lemonade	Others
Boys	2	8	12	2	6
Girls	4	10	9	3	4
Total frequency	6	18	21	5	10

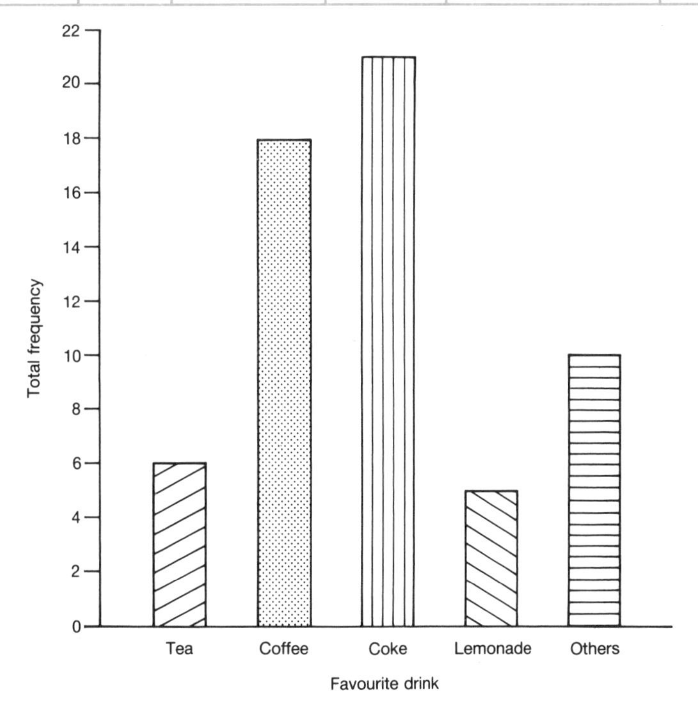

It is usual to separate out each bar by a space of the same width as the bars. Care should be taken in selecting a suitable scale for the total frequency so that your data is clearly displayed.

The same information could be shown on the same axes but showing the boys' preferences separately from the girls'. By doing this you are able to compare not only the types of drinks favoured by a class, but also how those preferences are broken down between boys and girls.

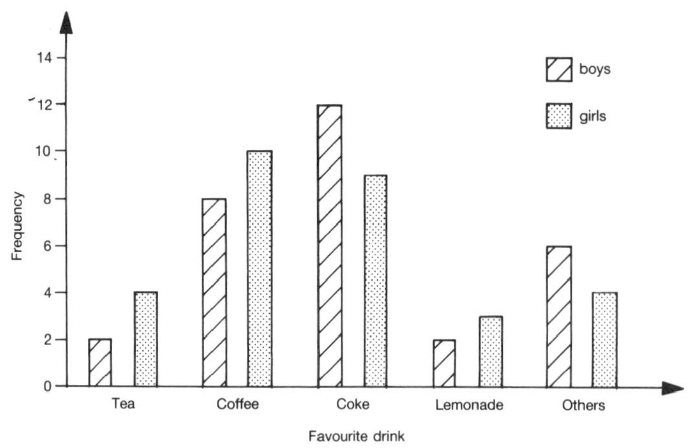

Bar charts are the easiest of the diagrams to draw and to understand. They show effectively how one section compares with another within your collection of data. However, unlike a pie chart, they do not give much idea of how the *size* of one particular section compares with the *whole* group.

With slight modifications, bar charts can become vertical line graphs, histograms or block graphs, and these are explained in various textbooks.

Line graph

Line graphs show data by means of a drawn line joining items of data and are ideal for showing upward or downward trends. Usually, one of the variables will be the passing of time. Examples of trends that can be shown this way would be:

(a) the temperature of a hospital patient

(b) the amount of rainfall per month

(c) sales figures for any kind of business

(d) traffic surveys throughout the day.

A garage measured how many litres of petrol it sold each day during one week. Below is the table of sales and the line graph plotted from these values.

Day	Monday	Tuesday	Wednesday	Thursday	Friday	Saturday	Sunday
Litres	5000	4500	5500	4000	7000	8500	6500

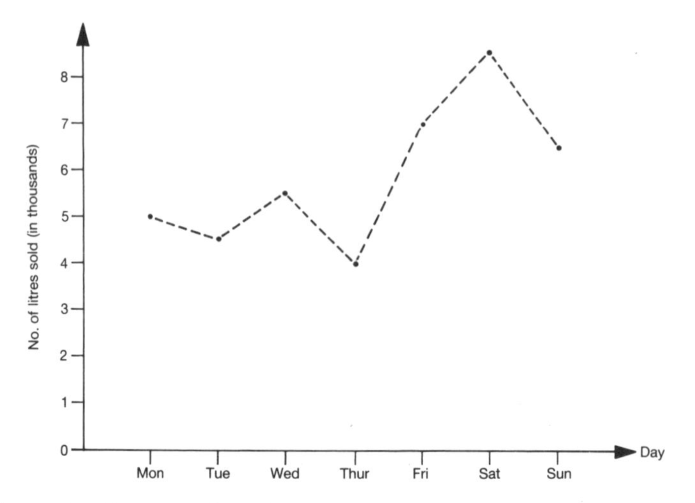

Notice how the line graph indicates a rise and fall in sales. You can spot immediately that business is steady during the early part of the week but has a sudden upsurge as the weekend approaches. This information would be valuable to the garage owner.

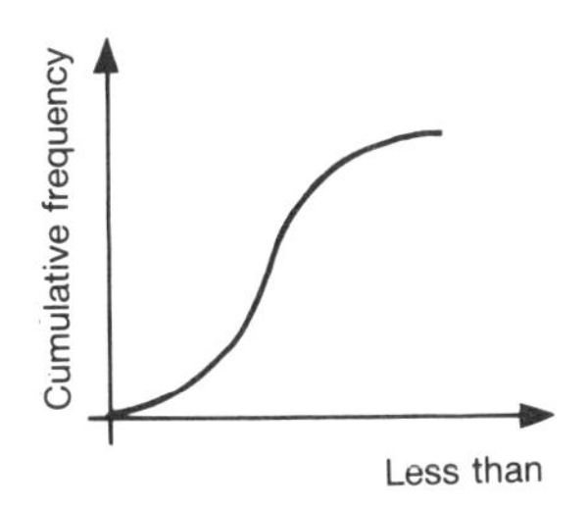

Cumulative frequency diagram

When you draw a graph of a cumulative frequency distribution, such as the distribution shown on page 46, you will get a *characteristic shape* that is called a **cumulative frequency diagram**. This shape is indicated here.

Sometimes, as shown, the points are joined by a smooth curve. Often, however, the points are joined by short straight lines from one point to the next. The choice is yours, but drawing smooth curves does require more practice. In the first instance you will find it neater and easier if you join successive points by a straight line.

To construct a cumulative frequency diagram, follow these simple instructions.

(a) Draw two axes on graph paper.

(b) Mark off the vertical axis, to a suitable scale, as your cumulative frequency.

(c) Mark off the horizontal axis to a suitable scale, plotting the points as 'less than' whatever units you are using (e.g. Marks less than).

(d) Plot each of the cumulative frequencies against its corresponding 'less than' value.

(e) Join up the points plotted by a smooth curve or by short straight lines.

Mrs Jones works out the cumulative frequency distribution of the 5th year mock examination results in mathematics and then draws the cumulative frequency diagram. There were 120 candidates.

Marks (less than)	10	20	30	40	50	60	70	80	90	100
Cumulative frequency	0	2	14	30	48	72	102	110	118	120

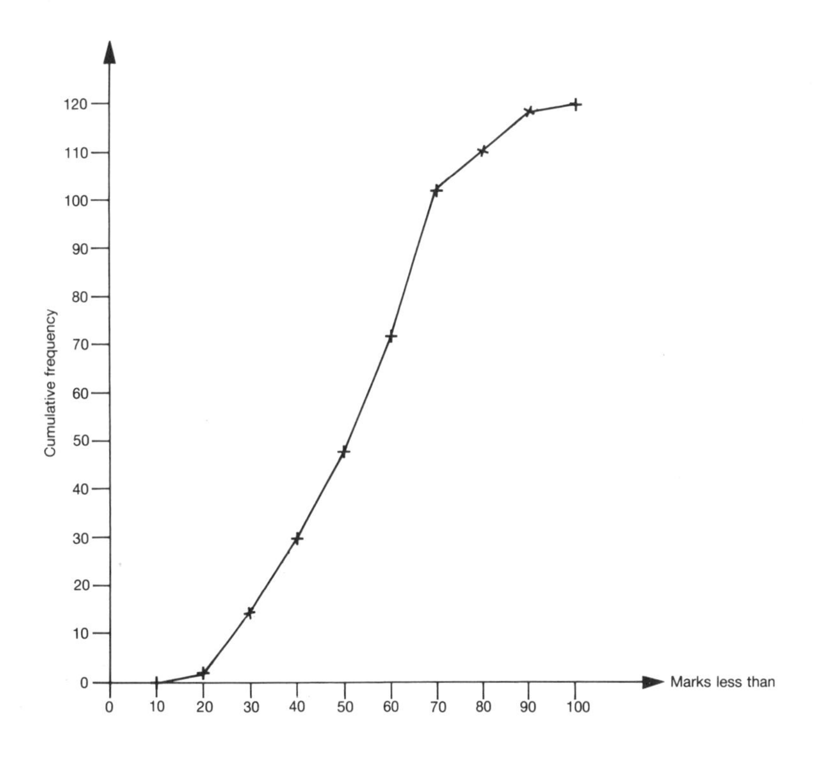

An advantage of this type of diagram is that it shows the **spread** of results.

Warning

You may not be able to use all of these types of pictorial representation to illustrate your data. Do, however, use *as many as possible* and don't stick to one particular type.

Averages

Refer back to the flow chart on page 40 to see where averages fit into the design of your survey.

You have tabulated your data so that it can be easily understood, but a value that characterizes the whole group and is easier to imagine is called the **average** of the group.

There are three averages used in statistics:

(a) mode

(b) arithmetic mean

(c) median.

Averages are usually near to the centre of the data and are often called **measures of central tendency**.

The **mode** is the value that *occurs most often*.

The **arithmetic mean**, often called the **mean** is found by calculation:

$$\text{arithmetic mean} = \frac{\text{sum of all the values}}{\text{total number of values}}$$

The **median** is the *middle value*, but only when all the numbers are arranged in *order of size*. If there is no middle value then take the half-way point between the two middle values.

Each one of the three averages has its own advantages and disadvantages and its own applications. For example fashion shops will be most interested in the *mode* i.e. the most fashionable – most frequent – most common size.

During the first ten matches of the season, the goals scored by the school team were:

0, 0, 1, 1, 1, 2, 2, 3, 3, 5

Mode = 1 goal (the score occurring most often)

Mean $= \frac{0+0+1+1+1+2+2+3+3+5}{10} = \frac{18}{10} = 1.8$ goals

Median = 1.5 (half-way between 1 and 2, the two middle values)

A die is thrown 20 times and the results are tabulated as below.

Score on die	1	2	3	4	5	6
Frequency	2	4	2	5	4	3

Mode = 4 (scored on five occasions)

Mean $= \frac{(1\times2)+(2\times4)+(3\times2)+(4\times5)+(5\times4)+(6\times3)}{20} = \frac{74}{20} = 3.7$

Median = 4 (the middle score is one of the five occasions that 4 was scored)

A small factory employs 50 people whose weekly wages are shown in the table below. (Here we are working with groups.)

Wages (£)	Frequency	Mid-point (£)
100-119	6	109.50
120-139	11	129.50
140-159	22	149.50
160-179	7	169.50
180-200	4	190.00

Here we can identify the most common *group*, which is the **modal class**.

Modal class = £140-£159

Grouped mean

$$= \frac{(109.50 \times 6) + (129.50 \times 11) + (149.50 \times 22) + (169.50 \times 7) + (190 \times 4)}{50}$$

= £146.34

(For each *class* (i.e. 100-119, 120-139 etc) multiply the frequency by the mid-point of the upper class limit and the lower class limit, when calculating the grouped mean.)

The **median** is the value half-way between the wages of the eighth and ninth lowest paid workers in the £140-159 class. Since we do not have individual wages listed, this is as near as we can get to giving a value for the median. (Sometimes we may make a guess of 'about £147'.)

There are further calculations that can be done using 'guessed means', 'weighted means' and 'moving averages' which can be easily researched from textbooks. Generally, these aspects would be used only by higher level candidates.

Spread of results

We mentioned earlier the **range** of values obtained when collecting raw data. This gives us an idea of the measure of dispersion – the **spread** of the results.

Rather than just using an average to describe a set of results, it is always better to measure the spread of the results as well.

Two sets of data may have the same average value but be entirely different in the way they are spread out about that average value.

For example,
Set 1 = {4, 5, 5, 5, 6} Set 2 = {1, 3, 5, 7, 9}
Set 1 has exactly the same mean (average) as Set 2 but the spread of data is different.

Two important measures of dispersion are:

(a) range

(b) interquartile range

Range

The **range** is the difference between the highest and lowest values in the data.

Interquartile range

If the data is put in numerical order, then the **interquartile range** is the difference between a quarter of the way up the range and three quarters of the way up the range.

FULL SET OF DATA IN ORDER		
↑	↑	↑
Lower quartile	Median	Upper quartile

Interquartile range = upper quartile – lower quartile

Interpretation of results

In the very beginning, it is helpful if you can state a hypothesis such as:

'A pedestrian crossing is needed outside our school'

By doing this you are setting yourself a target that you can test and possibly justify by your research. (This is much better than, say, simply conducting a traffic survey without any clear aim in mind.) This method will help you to structure your assignment and enable you to draw conclusions. There has to be an initial hypothesis and in the end, conclusions drawn, otherwise you will have conducted your survey with little point to the exercise.

Interpretation of your results could follow from the graphs, the averages, and/or the spread of the results. You have to recognize 'key' results which either support or work against your initial hypothesis. Some results might help you to make predictions or comparisons.

You should mention the limitations of your survey i.e. lack of numbers in survey, lack of time to collect more data etc, and say what could have been improved given more time and data.

Most important, however, is the way in which you draw your valid conclusions together and make an accurate statement about them.

Communication of results – presentation

You must try to present a clear report which is well structured and well written. You have to put forward your arguments, backed up by the results of your survey. Here is your **checklist** which should form the skeleton of your survey.

(a) Hypothesis.

(b) Collection of raw data.

(c) Tabulation of data.

(d) Pictorial representation (use plenty of colour).

(e) Averages (where appropriate).

(f) Spread of results (where appropriate).

(g) Interpretation of results.

(h) Have you proved what you set out to prove?
Say whether or not in your concluding paragraph.

If your work is inconclusive then state as much, but qualify your statement by suggesting further work which could be undertaken in order to reach a conclusion.

Probability

Probability could be an area that appeals to you in a statistical assignment. Phrases such as 'dead cert', 'no chance', 'odds on' or 'fifty-fifty' are common language today. At some stage during your life you will gamble, not necessarily with money at the bookmakers, but on an event happening or not happening, as the case may be. When you go outdoors without a raincoat you are taking a chance that it will not rain.

The theory behind probability is most interesting and much can be learned by carrying out probability experiments. As one of your assignments you might like to carry out such experiments, but bear in mind you will have to be able to draw conclusions from your results.

There are plenty of experiments that can be performed with dice, playing cards, spinners, drawing pins, coins etc, which will help you to understand the theory of probability.

A simple statistics/probability assignment might begin with a statement, which is in fact two hypotheses:

> 'Football is the most popular sport amongst the boys in our school. If I were to ask *any* boy in our school what his favourite sport was, the *probability* that he would choose football would be more than half.'

You could then conduct a survey of *all* the boys in your school and then perform a probability experiment by asking boys *at random* their favourite sport. Having done both of these you could either support your original hypotheses or amend them in the light of your results.

Before you start, make out a plan similar to that outlined below and submit your plan as part of your assignment.

STATISTICS PLAN

Name ______________________ **Form** ________

Teacher ______________________ **Date** ________

1 Hypothesis

2 Are you going to work with others?

3 How are you going to collect your data?

4 Do you need to go out of school?

5 Do you need money? (bus fares etc)

6 How long will it take to collect the data?

7 Will you need practical equipment? If so, what?

8 What will you do with the data you collect?

9 Give an outline of what you imagine the final assignment will look like:

10 Identify the skills you are going to use:

(a) tabulation

(b) diagrams

(c) averages

(d) spread of results

Suggested areas for study

Car survey (e.g. make, age, colour, car parking etc)
TV survey
Reaction times – ITMA Shell Centre has a computer program
Weather
Sports Results
Opinion Polls
Advertising Claims
Length of phone calls from a phone box
School library
Words in a book
Expiry dates on food
Accidents
Leisure Time
Class statistics
Journey times
Shop prices
DIY
Adverts on TV and in newspapers
School Meals

This list is by no means exhaustive. If you have any other areas for study then discuss them with your teacher.

Sample assignment

Multi-level

'Heights of boys and girls in your class'

- Collection of data
- Classification and tabulation
- Frequency distribution and cumulative frequency tables
- Averages
- Spread of results

PLAN

(a) Take whatever measurements you will need. Explain exactly how you did this.

(b) Arrange the frequency tables in three sections:

(i) boys **(ii)** girls **(iii)** all together

(c) Calculate the cumulative frequency table.

(d) Use your distribution to work out modes, means, medians, range, interquartile range etc.

(e) Describe how your results would be affected if your class had been joined by a very tall girl and/or a very short boy.

(f) Describe how your results would compare with a class:

(i) one year older **(ii)** one year younger

(g) Distinguish between all your results, especially the averages and ranges.

(h) Comment on the usefulness of your results for:

(i) a local clothes shop **(ii)** school

SECTION FOUR

Practical mathematics

What is 'practical mathematics'?

To describe someone as a **practical** person means that the person is clever at doing and making things. The word 'practical' suggests an activity, as distinct from study or theory, and it is also used to describe something that is useful.

Practical mathematics would therefore suggest:

(a) an element of making things

(b) an application of mathematics that is useful

(c) mathematics that involves an activity.

It follows, therefore, that mathematical surveys or probability experiments are examples of 'practical mathematics' because they involve activities.

Practical geometry

We shall concentrate in this Section on **practical geometry** which appears in the requirements of a number of examination boards.

Geometry is the study of lines, angles, surfaces and solids. Practical geometry is best regarded as an activity, everyday application or modelling that involves:

(a) lines

(b) angles

(c) surfaces

(d) solids

This gives you a wide variety of assignments that you could undertake but, before you start, make sure that your teacher approves of what you are doing. Follow the familiar pattern of making a **plan** for your assignment and keeping your **checklist** close by you for reference.

The Midlands Examining Group make some suggestions in the practical geometry category from which suitable topics may be chosen. However, you need not feel restricted to just their ideas. An original idea has more chance of scoring good marks, provided that it is done well. Their list includes:

1 simple surveying

2 scale drawing, maps

3 model making

4 construction of curves

5 spirals

6 geometry in nature

7 packaging

8 wheels and gears

Some of you may enjoy the practical aspect of say, model-making, as you will have the opportunity to make things as well as writing up your work and doing calculations. To many pupils, model-making means making solid shapes from nets, or using straws. However, there are specially designed construction kits that your school may have, so ask your teacher if he or she can supply you with items that you will need.

Rather than simply making solid shapes from card, although this has some merit as an assignment in itself, you might try the idea of **dissecting** shapes, that is, cutting them up in some special way with a slice or maybe two slices.

Dissecting shapes

'Take a basic solid shape or shapes and investigate the parts obtained when you slice through your basic shapes in some way.'

The practical equipment you will need for this is:

cardboard	protractors	shape template
scissors	sharp cutters	coloured pencils
rulers	glue	set squares
pencils	polystyrene	coloured paper
compasses	calculator	Blu Tack

'It might be a good idea to get your woodcraft teacher to help you make up shapes and then make the necessary slices using suitable machinery. You could cut a cube made out of polystyrene to work out the nets in this assignment.'

Look carefully at the way the **cube** is used as my basic solid shape. This could easily be adapted to cuboids, pyramids, regular prisms etc.

Cube

We shall look at three examples and see how you can investigate them.

Easiest slice ***Not quite so easy*** ***Most difficult (not as easy as it seems)***

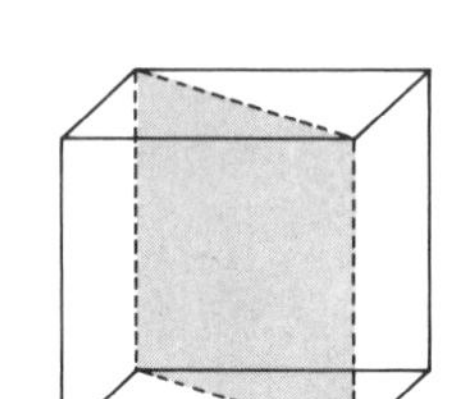

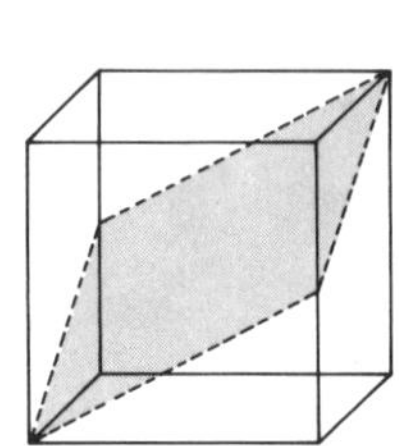

The easiest slice

For this, the identical halves can be made from this **net**.

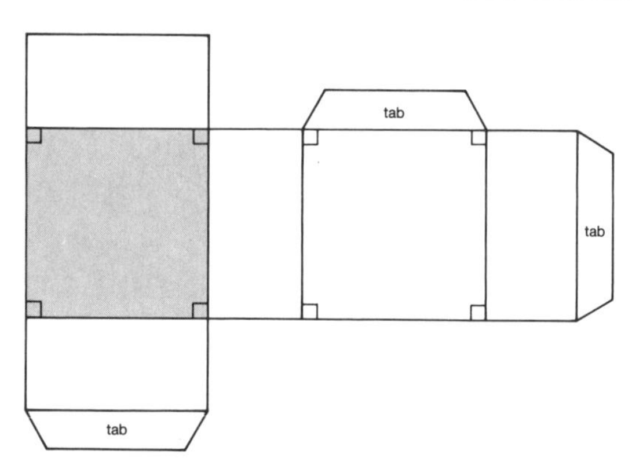

All angles are 90°. Do not forget your **tabs** if you are making the model.

The not quite so easy slice

For this, the identical halves can be made from this net.

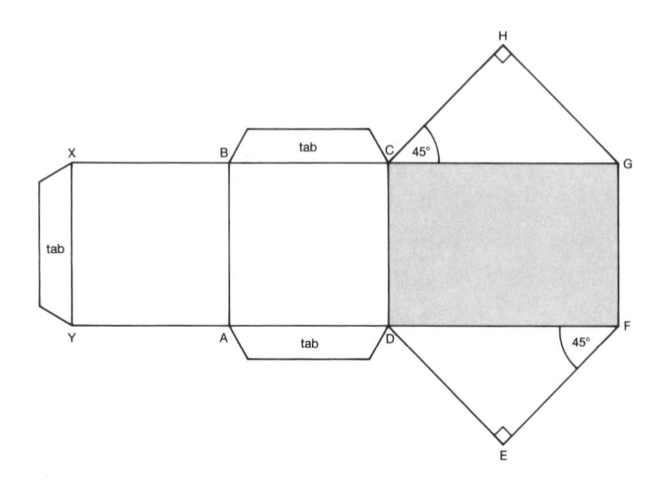

The length of the side ED is the same as the length AD.

Similarly, the length of CH is the same as the length BC.

Most difficult slice

For the most difficult, the net would look like this:

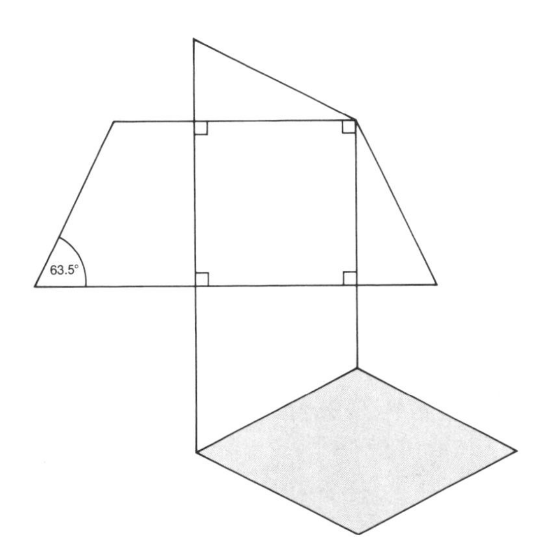

Other shapes

Dissecting is an activity that you can practise and, as you get more and more proficient, you can start to dissect your solids into more than two shapes.

Here are some other ideas for other basic shapes, with the slices shown by the dashed lines.

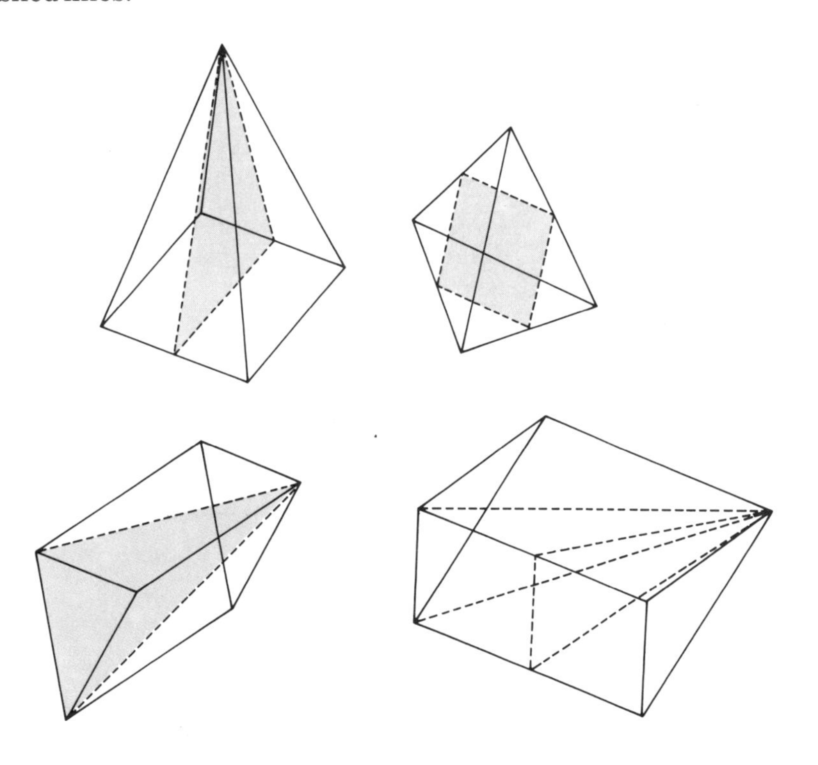

Your assignment

Your assignment should include a written account of how the nets were designed as well as the final models. The final account should include accurate drawings of the nets used for the models and, most importantly, explanations of all your work. All calculations must be shown.

To give you some idea of how to plan your study over a period of three to four weeks, follow the next assignment on **equable shapes**. Each stage is carefully planned out, developing further what you have already found out. Your aim should be to get through as many stages as is possible.

Equable shapes

This assignment is suitable for all levels

This topic is about finding two-dimensional shapes in which the **area** is numerically equal to the **perimeter**.

It extends to include three-dimensional shapes in which the **total surface area** is numerically equal to the **volume**. The variety of shapes will depend on which level the candidate is entered for.

The assignment requires an **investigational** approach. The mathematics involved is:

1 area and perimeter calculations

2 areas of triangles etc

3 volumes and total surface areas of cubes

4 further areas, surface areas and volumes concerning circles, cylinders, spheres

5 forming linear equations

6 solving linear equations

7 Pythagoras' theorem

8 solving quadratic equations (Higher level only)

The resources needed are:

squared paper	pens and pencils	calculator
plain paper	protractor	cubes
rulers	compasses	computer

What now follows is a stage-by-stage guide through the assignment. You should work your way through, as far as possible.

Stage 1

On squared paper, experiment with various rectangles, trying to find any in which the area is numerically equal to the perimeter.

Whole-number solutions will give you two rectangles, as shown in the diagram.

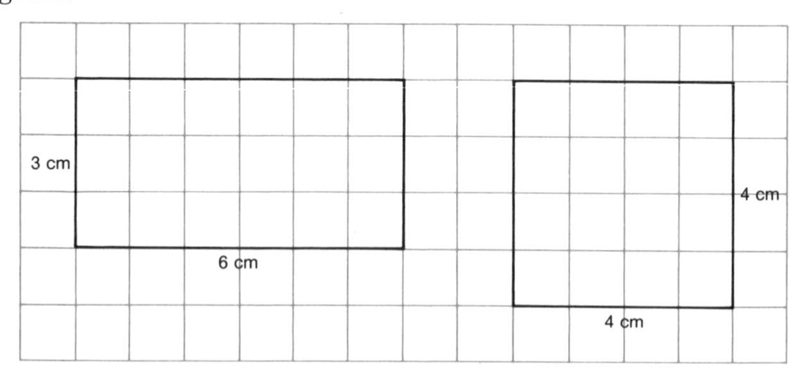

Stage 2

Proceed further with rectangles by 'adding squares' to them.

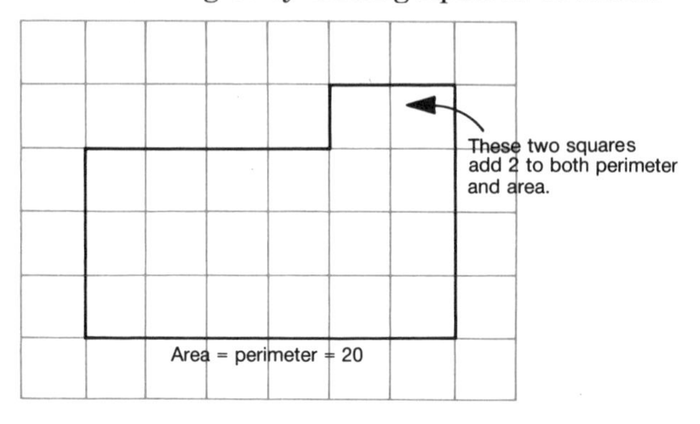

Explore different ways of adding squares. Record your results. Are there any area/perimeter values that cannot be obtained?

Stage 3

Try to find values for the length and width of a rectangle that are not whole numbers, yet they give the same value for area and perimeter.

Start with a length of 5 cm and try to work out an appropriate width. This is possible by trial and error using a calculator.

Stage 4 Use of algebra

Try to solve the problem in Stage 3, not by trial and error, but by the use of algebra.

Form an equation for area = perimeter, and solve it.

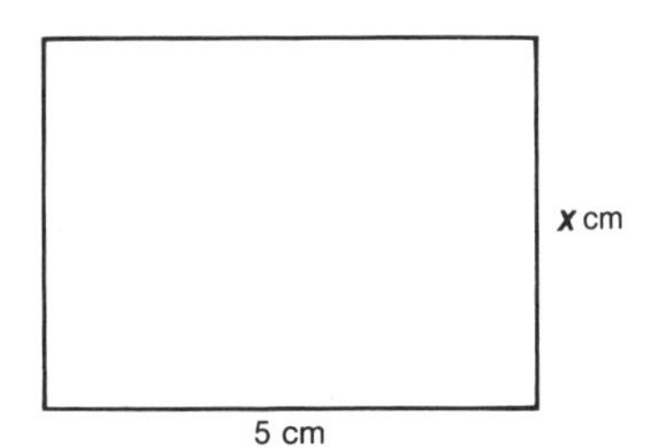

Here, area $= 5x$
perimeter $= 10 + 2x$

Making $5x = 10 + 2x$
$3x = 10$
$x = 3\frac{1}{3}$

Complete the table for other values of length.

Length cm	1	2	3	4	5	6	7	8	9	10
Width (x) cm			6	4	$3\frac{1}{3}$	3				

Plot a graph of 'length (L) against width (W)' for those rectangles where the area and perimeter have equal numerical value.

Re-arrange your formula $W \times L = 2(W + L)$ making W the subject of the formula.

Stage 5

Extend the problem to three-dimensions. Try to find the dimensions of a cube for which the volume is equal to the surface area.

A solution can be found by trial and error. Another way is by solving the equation $x^3 = 6x^2$ where x is the side length of the cube.

Now try to find solutions with cuboids instead of cubes. There are quite a few whole-number solutions to your problem.

Stage 6

As with Stage 2 for squares, now try 'adding cubes'. You will find this is a more difficult problem in three-dimensions.

Stage 7

Extend the problem to **equable triangles**. Start with right-angled triangles and find which have numerically equal areas and perimeters. An understanding of Pythagoras' theorem is essential here.

Solve the problem:

> 'What right-angled triangle of base 5 cm has numerically equal area and perimeter?'

Then proceed to explore other kinds of triangles. Try equilateral triangles, if you can.

Trial and error solutions are possible, as are solutions using algebra.

Stage 8

Extend the problems to circles. Can you find the area and perimeter of a circle of radius 5 cm?

If **area** = **circumference** then it is easy to see that $\pi r^2 = 2\pi r$ which gives $r = 2$ as a solution, where r is the radius.

This fact leads to an interesting development. Draw circles of radius 2 cm and divide them into equal parts.

Construct regular polygons inside these. What do you think happens with the area and perimeter of these polygons?

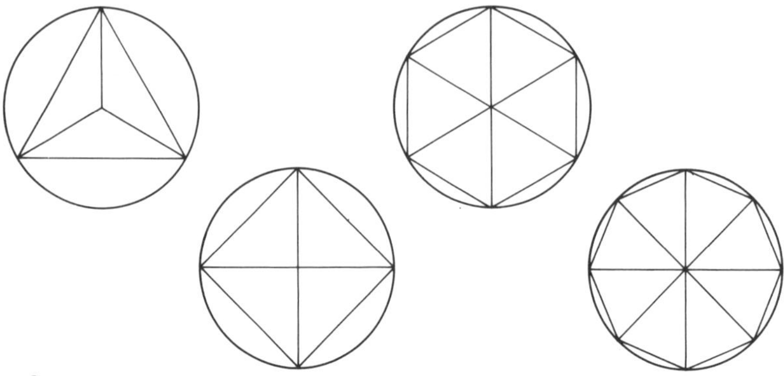

Stage 9

Extensions to the problems are possible, such as:

(a) Find the dimensions of *solid cylinders* that have the 'volume = surface area' property.

(b) Find the dimensions of the *sphere* that has the 'volume = surface area' property.

(c) Find other shapes involving circles which have the 'area = perimeter' property, e.g.

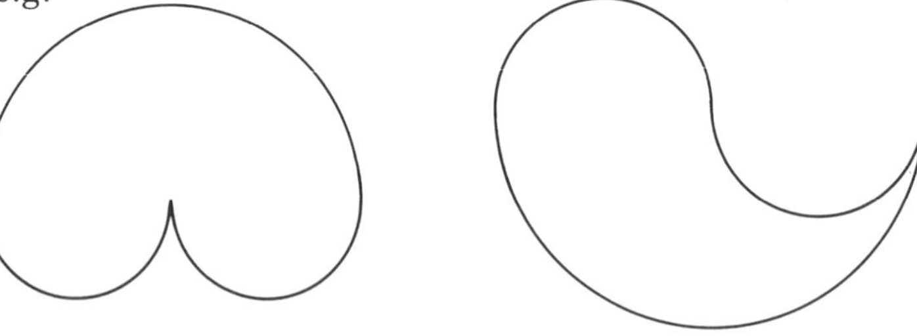

(d) In Stage 8, an interesting fact emerged about the circle that circumscribes regular polygons. Does a similar thing happen with a sphere and regular solid shapes e.g. cubes, tetrahedra etc?

Everything stops for tea

This is an example of **practical work** given by the Southern Examining Group. It is intended to enable you to understand concepts e.g. area and volume etc and to provide the opportunity for you to measure things for a purpose. You would be expected to have done some other form of practical work before going ahead with this assignment.

Task

A tea firm wants to make presentation boxes of 200 g of tea. The card used to make the boxes costs 20p per square metre. Make three different boxes the firm could use. Write a report for the Managing Director to explain your conclusions.

Points looked for in marking

Comprehension

- Understanding of need for boxes of set volume.
- What volume is 200 g of tea?
- What does 'different' mean? Are two cuboids different?
- Are all boxes suitable? (e.g. tetrahedra?)

Planning

- Find necessary volume. Choice of method.
- Find size of boxes.
- Make boxes.
- Find costings.
- Write report. (Keep notes as you go along.)

Carrying out the task

- Are the boxes produced suitable? (One a little over the required volume might be satisfactory, but one a little under would be unacceptable.)
- How accurate were the measurements?
- What checks were made?
- Were the sizes of the boxes found by trial and error or theoretically?
- Which came first, the net or the box?
- Is allowance made for tabs etc?
- Is wastage considered in the costing?

Communication

- Can the assessor know and understand what was done at each stage?
- Does the report for the Managing Director bring out the important points?
- Is there any comparison made between a 'practical box' and an 'attractive box'?
- Oral work could bring out a consideration of determining which shape would be most acceptable to the public, and how this would affect the costings.

Follow-up work

The following areas of enquiry could be suitable for practical work:

1. modelling polyhedra etc
2. measuring heights and weights e.g. of boys and girls
3. experimental determination of volumes e.g. by immersion
4. gradients, using a clinometer etc
5. finding heights of buildings
6. pendulums
7. conic sections
8. tessellations and patterns
9. weighing and density
10. speed, distance and time

Although the Group describes its list as simply 'practical mathematics', only items 2, 9 and 10 would not be suitable for a 'practical geometry' assignment.

A student's assignment

Finally in this section, is a student's assignment entitled 'Open Boxes', together with the teacher's comments.

Teacher's comments

Open Boxes

Shows good understanding of the task

The open boxes were made by cutting a square out of each corner of a 30 × 20 inch rectangular piece of card. The four outside edges were folded upwards to form the sides of the box.

Good diagram to illustrate practical work.

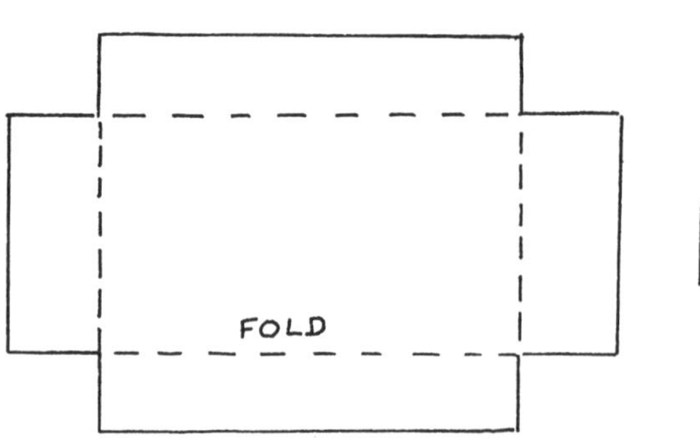

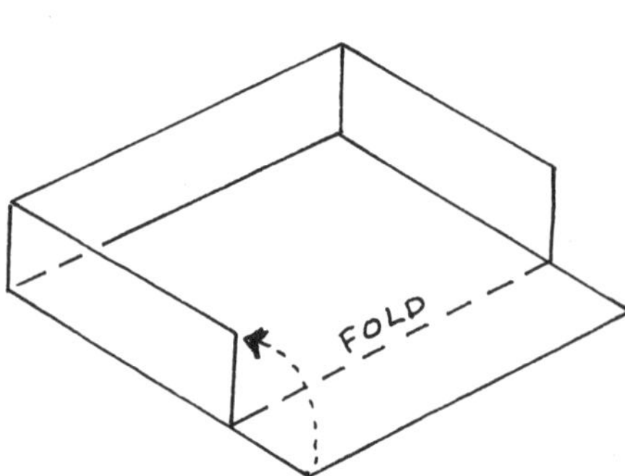

I made several boxes like this but each time made the size of the square that I cut out from each corner different.

I basically ended up with three different types of boxes (cuboids).

a) shallow boxes with large base area

b) tall narrow boxes with small base area

c) boxes approaching a cube in shape

I like the idea of the predictions. You have given a purpose for the assignment, which you can either justify or not.

I would imagine that the shallow ones would have a greater volume and therefore hold more. If you were making a lot of these boxes you would be interested in the most economical box (the one that uses the least card). I predict that the boxes approaching a cube in shape would use the least card.

Each box was made and labelled.

Box 1 had corners cut out 1 × 1 inch

Box 2 had corners cut out 2 × 2 inch

Box 3 had corners cut out 3 × 3 inch

and so on.

Teacher's comments

A good choice of strategy.
(i) be systematic
(ii) make a table
(iii) spot patterns

Accuracy very good.

Solid summary of results.

No attempt made to use algebra! Is it possible?

Vague generalizations.

Even so, submit the models.

I should like to see more of this.

Results

Box number	Length	Width	Depth	Volume	Surface area
1	28	18	1	504	596
2	26	16	2	832	584
3	24	14	3	1008	564
4	22	12	4	1056	536
5	20	10	5	1000	500
6	18	8	6	864	456
7	16	6	7	672	404
8	14	4	8	448	344
9	12	2	9	216	276
	inches	inches	inches	inches^3	inches^2

The pattern of numbers for the volume suggests shallow boxes have smaller volumes gradually rising to cube type boxes having a maximum volume and then decreasing with very tall boxes again having small volumes.

The pattern was slightly different for the surface area. The shallow boxes had the largest surface area which gradually decreased until we reached the very tall boxes which had the smallest surface area.

I could have arrived at this result without making every box.
The box that had the largest volume was box 4 (22 × 12 × 4 inches).

What happens if . . . ?

Teacher's comments

This could have been arrived at if you had drawn a graph to represent the table.

a) I cut out corners measuring 3·5 x 3·5 inches.

Volume = 3·5 x 23 x 13
= 1046·5 inches^3

b) I cut out corners measuring 4·5 x 4·5 inches.

Volume = 4·5 x 21 x 11
= 1039·5 inches^3

c) I cut out corners measuring 3·9 x 3·9 inches.

Volume = 3·9 x 22·2 x 12·2
= 1056·276 inches^3

This would have been a suitable extension to the problem

There is a box that will hold more than box 4 but the dimensions are awkward and impractical.

Conclusion

I like the way you are not afraid to test your conjectures – even when they are wrong.

My predictions were not very accurate. I suppose a 22 x 12 x 4 inches box could be described as a shallow box but it was one of the middle range.
I was totally wrong with the surface area. The taller the box the smaller the surface area.

This too is another extension. At least you have made the assignment 'real life' with your conclusions.

Boxes are used to hold things. They usually contain saleable items so we need the box that holds the most yet uses least amount of card unless the box is covered with advertising. In this case we would need a large face other than the base.

A good attempt at a 'short assignment'. I like the overall layout – your arguments are well presented. Although the mathematics was accurate, more could have been incorporated e.g. graphs, etc. Further extensions to the task needed to be explored for a higher grade.

SECTION FIVE

An extended piece of work

Everyday applications of mathematics

It is usual for assignments under the heading, 'Everyday applications of mathematics' to have a **theme** relating to common occurrences in life.

An extended piece of work has to be done over a lengthy period of time and should show how you can develop your ideas from one stage to another. The piece of work may take various forms, and could focus on:

- an investigative piece of work
- solving problems
- pursuing an in-depth study
- a simulation
- a project using a computer
- practical work

Again, there is no prescribed finishing point. Neither is there any definite time set for your study. Your teacher will probably give you a deadline for handing it in, but your work *must* be sufficient for you to display:

(a) a sustained effort

(b) the development of mathematics beyond a single stage.

Your approach

It is extremely important that you know how to approach an extended piece of work. Here you have the opportunity to write a piece of work of your own choice, with the approval of your teacher, that will contribute a large proportion towards your final coursework mark.

An **extended piece of work** is 'open-ended', so you are free to follow your own ideas. You will be asked to investigate a topic which interests you and to make a written report of your findings. The purpose is as follows.

1 To allow you to show that you are capable of working for a sustained period of time to develop a substantial piece of work.

2 To enable you to show, through your topic, that you can apply the mathematical skills that you have acquired.

3 To enable you to show your ability to overcome difficulties.

Choosing a theme

Students can choose the theme of the work in a number of ways. The topic may arise out of an interest in other subjects at school, or a hobby, or a follow-up to a piece of work already studied. It may develop out of an activity at home, or have a purpose in solving some particular problem. However, for you to work to the best of your ability, your choice must hold your *interest* and motivate you over quite a long period of time.

Getting started

Your school will possibly give you about three weeks of lesson time plus homework (in some schools) to complete your extended piece of work. About a fortnight before the allotted time begins, your teacher should suggest that you start to think about a suitable title.

Your teacher will outline what is required by the examination group, offering help and advice. He or she will try to give you some ideas for suitable titles, but you should have an open mind at this stage.

Go home and talk things over with a parent or older brothers and sisters if you have any. You might need them to take you to places or to collect things for you – so stay in their good books!

Remember – you must be interested in your choice of title if you are to do a good piece of work.

Choosing a title

Let us assume that you have chosen a subject area that interests you and will give you the opportunity to express yourself.

Now choose a **meaningful title**.

> 'THE DIABETIC CHILD' – is an example of a *vague* title.
>
> 'THE ADDED COST OF THE DIET FOR A DIABETIC CHILD' – is much more *meaningful*.

Planning the extended piece of work

This is a very important stage of a good piece of work and in some cases you may be asked to submit your **plan** as part of the assessment. Write down questions that you need to answer concerning your work.

1 How do I intend to start?

2 What information do I need?

3 What materials do I need?

4 Do I need to write to someone or even visit them?

5 What reference books do I need?

6 How do I intend to collect the information that I need?

7 How long will it take me to collect the information together?

8 Can I do all this within my own locality or will I need to travel any distance?

9 Will my piece of work cost me any money and can I afford it?

As you answer these questions you will begin to get a good idea as to whether your selection will be worthwhile and possible. Do not hesitate to discuss the answers to these questions with your teacher. He or she will probably be able to suggest ways of solving problems or put forward other aspects that you ought to consider before you complete your plans.

Think twice before throwing away any catalogues, holiday brochures, train time-tables and advertising leaflets. They could be useful as sources of information or inspiration.

Mathematical content

You have now reached the stage where you have to start thinking mathematically. You chosen theme must have the scope for quite a lot of mathematical content.

Remember – The mathematical content should be your *own work* and you must be prepared to explain your workings orally.

It might be a worthwhile exercise to write down some of the maths that you would expect to use in your topic. For example, supposing your choice for an extended piece of work was 'Designing a garden to be cost effective'. Then you might incorporate:

(a) area calculations

(b) perimeters of rectangles, circles and triangles

(c) costing of plants, fertilizer, etc

(d) cost comparison for different types of fencing

(e) tables, and show graphically the yields of vegetable produce.

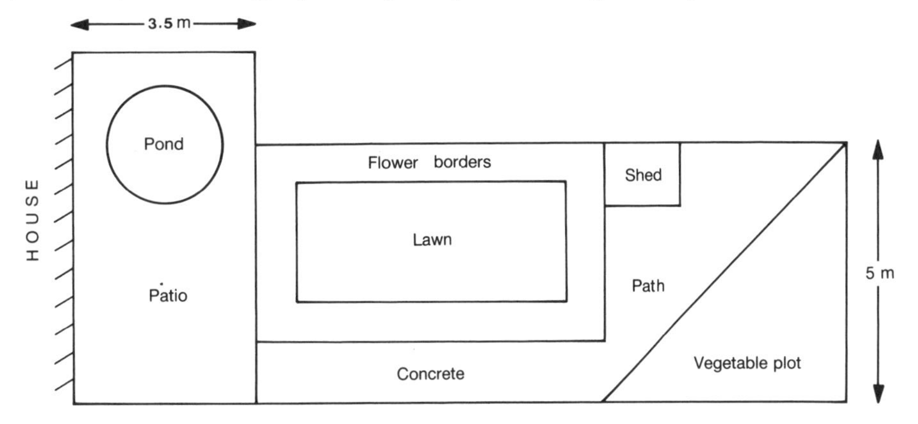

As you begin your basic calculations on the area of the vegetable plot, then you could extend this into:

'What different shapes will use up the same area?'

'Can I re-design my garden to give me a larger vegetable plot in the light of this calculation?'

Similarly, you could develop your perimeter calculations to lead into an exercise in surrounding the pond, shed etc with paving slabs.

The costing of plants and fertilizer could lead to a comparison between buying bedding plants or growing from seed.

With a fence around the vegetable plot, you could work out how the amount of light affected the yield of the crop by comparing it in different parts of the plot.

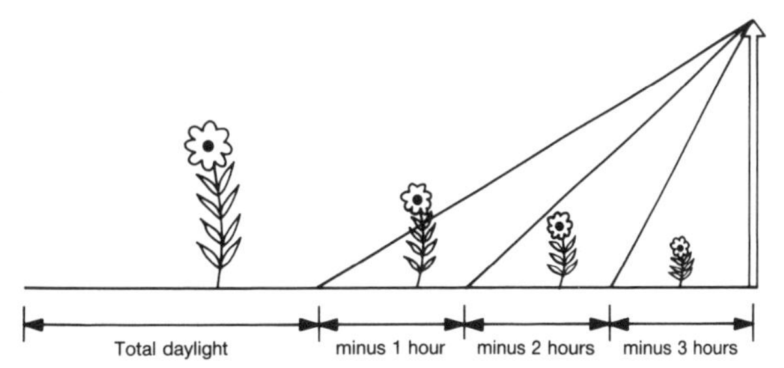

You must always remember the two basic requirements of a good maths topic, and ask yourself the following.

1 What sort of mathematics can I incorporate in the body of the work?

2 Can I build upon this to show the examiner that, as well as doing the basic calculations, I can develop some aspect further?

'Plan your work carefully and take notice of these warnings. They highlight the most common faults with coursework.'

Warnings

(a) Avoid doing page after page of writing without much evidence of mathematics.

(b) Avoid repetition of the same type of calculation.

Your finished work should contain a wide variety of mathematics including arithmetical calculations, diagrams, graphs, pie charts and tables. You should be clear in your mind *how* to present these. You will *not* be expected to perform calculations that your teacher has not covered, although there is no reason why such calculations cannot be attempted.

Do not hesitate to ask for help from your teacher.

Plan to be careful with your printing and layout. Neatness and good general appearance will create an impression with the person marking your work. It would be a good idea to do your finished article on file paper. This would make it easier to correct mistakes, to re-arrange sections or to repeat work should it be necessary.

A common fault is to do calculations just to create a good impression. By doing this you will only confuse the person reading your work. All your calculations should be relevant to the topic. Remember you will need to check all your work for accuracy; if you make mistakes you will lose marks.

We have spent some time dealing with the *importance* of the planning stage. You should discuss your plan with the teacher and, when approved, implement it. The amount of research, gathering of information and organization of the 'write-up' will depend upon your title, but once it is completed, you should be in a position to prepare for your finished article.

The write-up

You should be aware of the things that the marker is looking for in a good piece of work, apart from obvious things such as neatness and correct spelling. The basic guidelines for the award of marks are as follows.

1 Understanding the task

2 Planning

3 Mathematical content

4 Accuracy

5 Presentation and clear explanations

The extended piece of work is part of your learning process and help from the teacher is to be expected. Your mark will reflect your own personal contribution and the extent to which you are able to use the advice given to you by your teacher as the work develops.

Understanding the task

You should attempt to show:

(a) a clear understanding of the topic

(b) how appropriate mathematical problems can be extended

(c) clear reasons for doing the work in the way you have chosen

(d) appropriate use of skills

(e) how you can lay out your information

(f) how you can draw conclusions.

In your write-up, you should outline at the very beginning what you aim to achieve and in its final section, you should draw conclusions and state observations about your findings.

Mathematical content

Here, you have the opportunity to show the extent of your mathematical ability within a given theme. Your teacher will be looking for the following.

(a) The accurate processing of data.

(b) A clear discrimination between necessary calculations and what is irrelevant.

(c) A variety of mathematical tasks.

(d) A sound application of skills, knowledge and procedures to a problem.

(e) The making and testing of ideas and guesses.

(f) The formulating of general rules where applicable.

(g) The use of a range of diagrams and resources.

(h) Extensions to your line of thought.

Accuracy

It is vitally important that you are as accurate as you possibly can be in your calculations. Even if your calculations are simple arithmetic, you will be awarded marks for accuracy.

‘To write what appears on your calculator is another common fault. To be able to approximate is an important skill to learn and to show.’

If you are drawing lines, make sure you have a sharp pencil and that you draw a single line of the correct length. If you are measuring, make sure you have read your ruler correctly. When you are stating facts and figures, don't forget to put these in the correct **units**.

When you use a calculator be careful not to write down exactly what appears on the display. You have to be sensible and approximate to the *degree of accuracy* that you are working to, and that you require.

Significant figures

When working with approximate (*not exact*) measures, the general rule for the degree of accuracy of your answers is:

> When you multiply or divide, give the results to as *many significant figures* as there are in the *least exact* of the numbers you have used in the calculation.

My garden pond has a diameter of 2.85 metres, measured to *3 significant figures*. It is surrounded by crazy paving.

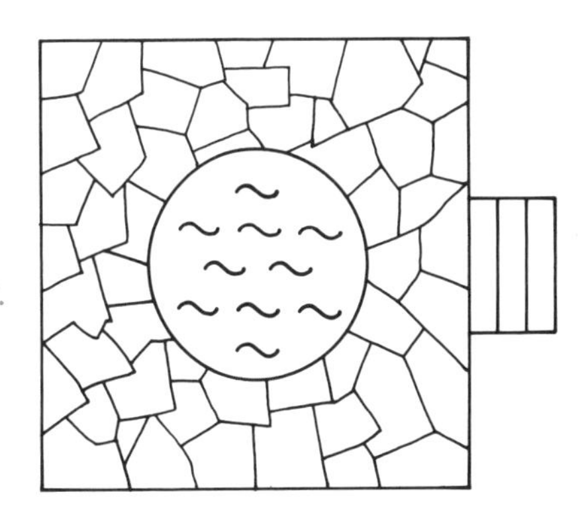

Circumference of pond $= \pi \times$ diameter
$= \pi \times 2.85$
$= 8.9535391$ (by calculator)
$= 8.95$ metres

(correct to 3 significant figures)

Area of pond $= \pi \times (\text{radius})^2$
$= \pi \times 1.425 \times 1.425$
$= 6.3793966$ (by calculator)
$= 6.38 \text{ metres}^2$

(correct to 3 significant figures)

Presentation and clarity

In your plan you will have stated what you hope to achieve. In your write-up you must state the results you have obtained and from them draw valid conclusions.

I am investigating the area of card needed to make cardboard boxes to put sweets in them. I have drawn two nets to show two different boxes.

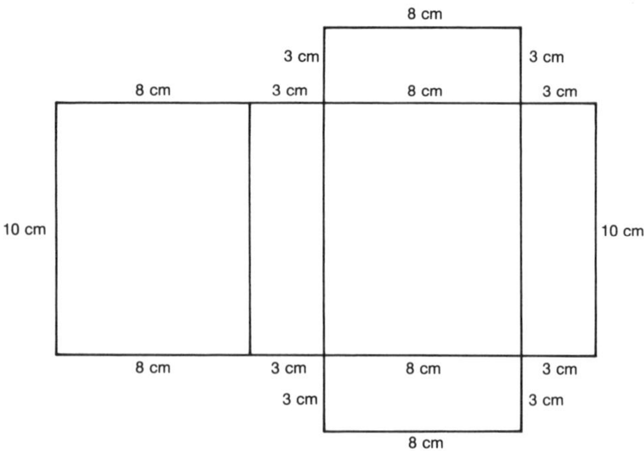

Total surface area of Box A

$$= (8 \times 3 \times 2) + (8 \times 10 \times 2) + (10 \times 3 \times 2)$$
$$= 48 + 160 + 60$$
$$= \underline{268 \text{ cm}^2}$$

Total volume $= 10 \times 8 \times 3$
$= \underline{240 \text{ cm}^3}$

Box B

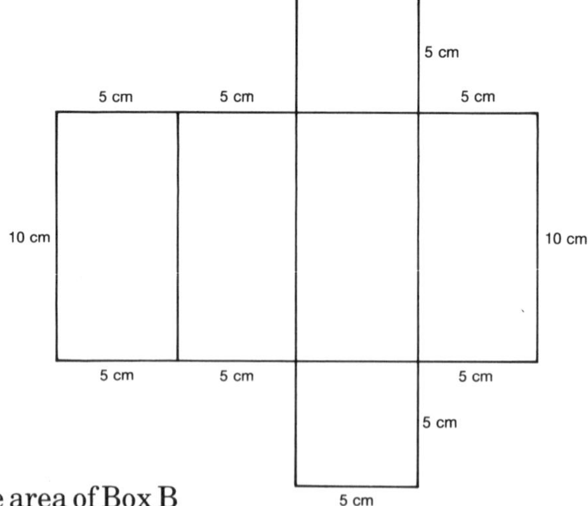

Total surface area of Box B

$$= (10 \times 5 \times 2) + (10 \times 5 \times 2) + (5 \times 5 \times 2)$$
$$= 100 + 100 + 50$$
$$= \underline{250 \text{ cm}^2}$$

Total volume $= 10 \times 5 \times 5$
$= \underline{250 \text{ cm}^3}$

'It is much easier to come to conclusions about your work if you have stated the aim of your assignment at the very beginning. Keep your aim in mind throughout your work. Have you done what you set out to achieve?'

Conclusion

Box B has a greater volume than Box A and will therefore hold more sweets.

Box B also uses less area of card than Box A and will therefore be more economical.

Box B is the better design for economical reasons. However, it may not be the best design for other reasons. For example, is it large enough to carry an attractive design? and can the firm fit its logo on the front of the box?

More on presentation and clarity

Your work must give a clear account of your aims, explaining any assumptions that you have made. This is another good reason for making a plan.

You should aim to communicate your results in the most appropriate way. Sometimes, a simple diagram can say more than a page of writing.

Also, your teacher will be looking for an effective use of a range of mathematical language and notation, diagrams, charts, and, where appropriate, computer output.

If your extended piece of work is about design, then produce a scale model.

At the very end of your work it is advisable to produce a **brief summary**. For this, ask yourself what conclusions you have reached, how much you have learned from your research, and whether you have discovered anything that you did not already know. You might state if your piece of work would help other students or people engaged in similar work.

An alternative approach

As an *alternative* to the plan of action that you have already been given in this Section, I offer the following.

1 Suggest a title.

2 Look at the syllabus to find what mathematics your assignment can illustrate.

3 Make out your project plan incorporating this list.

4 Use each area of mathematics in sections as you complete your write-up.

5 Draw conclusions.

'Your teacher will show you a list of mathematics that are in your syllabus.'

Some useful hints

'Everyday Applications of Mathematics' assignments are popular and, as a result, all tend to be somewhat alike. So, to end this section, there is a list of *Do's* and *Don'ts* to help you through *your* assignment.

Do's

- Make sure there is plenty of maths in your work.
- Try to be orginal.
- Explain your calculations.

Don'ts

- Don't copy from travel brochures or magazines.
- Don't paste in material from magazines etc, without describing what it means in your own words, and in relation to your work.
- Don't write too much.

There are practical skills that you can develop at home, which you might use in an everyday assignment. Why not practise these skills, which include:

- measuring lengths
- weighing objects
- estimating quantities (distance, time, mass, cost etc)
- scale drawing.

'You are surrounded by mathematics!! Be aware of the situations when mathematics are involved and skills can be practised.'

All of these skills could be used within a practical piece of coursework. Your home can be a source of realistic practical problems. You should find it easier to understand a practical problem if you can relate it to your own experience. Such problems might include:

- carpeting a bedroom
- carpeting a stairway
- using carpet squares
- decorating a room
- designing a kitchen
- deciding where radiators and pipes fit
- running a motorcycle or car
- fitting double glazing.

SECTION SIX

Oral assessment, aural tests and mental calculation

Oral assessment

In most assignments you will be assessed and awarded marks depending on how well you can **talk** about your chosen topic. This is another good reason for choosing something that interests you. Either during the assignment or immediately afterwards, your teacher may ask you questions related to the assignment. Although unusual, this may even take the form of an interview after you have completed your task, where your teacher will ask a set of questions about your work. You must be prepared to answer them clearly and as accurately as you can.

This is an area that you can practise before the event. Get a parent or friend to go through with you the list of questions on pages 74-75, so that you can try to give good answers. Why not use a tape recorder to hear for yourself how you are answering the questions? Listen to yourself and then see where you can improve your communication.

In the interview situation, try to relax, be positive and confident and use your own words.

I have listed some questions that you might be asked. Some may not be relevant to your own assignment, so try to pick out the most suitable for your needs. I have also tried to give you some idea as to what the teacher is looking for in your responses. You should note that I have *not* given the *answers* the teacher is looking for. My list gives, very briefly, the *areas* you could be thinking about when making your *own* answers.

Teacher's questions	Suggested areas for your answers (summarized)
Q 'What did you have to find out before you could get started?'	*Cost, transport, could I go out of school?*
Q 'Did you have to make any checks?'	*Was the library open? Were materials readily available? etc.*
Q 'How did you decide what sort of accuracy would be needed?'	*To the nearest penny, or most sensible approximation etc.*
Q 'What assumptions did you make?'	*Rooms rectangular, shapes symmetrical, buildings vertical etc.*
Q 'How did you record your information?'	*Tally charts, questionnaire, on squared paper etc.*

Teacher's questions	Suggested areas for your answers (summarized)
Q 'Why did you think this method was the best?'	*Clearer results, easy to read, to the right degree of accuracy etc.*
Q 'Did you consider other methods? If so, which?'	*Tables, charts etc.*
Q 'What information did you want your assignment to convey?'	*Based on title and objectives set out in the plan.*
Q 'How did you reach your results?'	*By calculation, by graphs etc.*
Q 'What do your results mean/show?'	*There is a need for . . . I have clearly shown that . . . etc.*
Q 'Were your results what you expected?'	*I was surprised by . . . I didn't realize that . . .*
Q 'Did anything else come out of your results?'	*Although I set out to show . . . it was also obvious that . . . etc.*
Q 'How did you check your work?'	*Discussion with parents or teacher; borrowed books etc.*
Q 'Did you have to correct anything?'	*Depends on results.*
Q 'Did you follow false leads?'	*Depends on results.*
Q 'Did you consider extensions to your work?'	*Depends on results.*
Q 'Did your assignment overlap into other branches of mathematics?'	*Depends on results.*
Q 'Did your assignment overlap into other subjects?'	*Could be science, geography, technology etc.*
Q 'Can you describe the most significant piece of knowledge you have gained?'	*Put things in order of importance when answering this.*
Q 'Are you a wiser person having done your assignment?'	*I didn't know that . . . I feel more confident when using . . . etc.*

Aural tests

There was a time when some emphasis was placed on mental arithmetic in most schools. In recent years, however, it seems to have gone out of 'fashion'. Although the calculator has taken over many of the numerical chores, there is still the need for you to do simple calculations, either in your head or using pencil and paper. It is essential that you are able to make **rough estimates** and have some idea as to how numbers behave.

Aural tests will not be purely mental arithmetic. Questions will be asked in a particular context. For example, '6 multiplied by 15' could be rephrased as, 'How much will 6 cakes cost at 15p each?'

You will be required to make quick estimates and approximations. For example, 'How many 18p stamps can I buy for £2.00?'

When doing these aural tests you may tend to rely on certain key words as **share** (= divide), **altogether** (= add) and **times** (= multiply). Some questions will deliberately use these keywords, yet the operation that you are expected to perform is entirely different, so listen carefully.

e.g. A group of six people *shared* some prize money. Each person got £30.00. How much prize money was there *altogether*?

Here you actually multiply 6 by £30.00, *not* divide or add as the key words might suggest.

There are some popular questions that you can do easily if you are aware of the correct techniques.

Example 1

What is the cost of 5 LPs at £5.99 each?

Cost = 5 × £5.99
= 5 × £6 less 5 × 1p
= £30.00 less 5p
= £29.95

Example 2

A garage bill is £66.00 plus VAT at 15%. What is the VAT?

15% of £66.00 = 10% of £66.00 + 5% of £66.00
10% of £66.00 = £6.60
5% of £66.00 = £3.30
So, 15% of £66.00 = £9.90

Example 3

Find 8% of £15.00.

8% = 8p in every £1.00
So, 8% of £15.00 = 15 × 8p = £1.20

‘Make up some questions of your own similar to these examples. These are skills to master – not only for your coursework but for everyday use.’

Example 4

A film starts at 7.25 p.m. and ends at 10.05 p.m. How long does the film last?

7.25 p.m. to 8.00 p.m. = 35 minutes
8.00 p.m. to 10.00 p.m. = 2 hours
10.00 p.m. to 10.05 p.m. = 5 minutes
So, 7.25 p.m. to 10.05 p.m. = 2 hours 40 minutes

Example 5

What change from £5.00 do I expect if I buy something for £2.63?

Do not try to take £2.63 from £5.00. Try counting from £2.63 until you reach £5.00.

£2.63 to £2.70 = 7p
£2.70 to £3.00 = 30p
£3.00 to £5.00 = £2.00
So, change = £2.37

Another technique to use when checking your answers is that of **approximations**. Always ask yourself, ‘Does my answer make sense?’

Example

How many passengers could be transported by a fleet of 27 buses each carrying 42 passengers?

42 × 27 = 1134 passengers

Check by approximation:
40 × 30 = 1200
My answer looks correct!

Example

On average, the sales of the best-selling single record during one particular week were 1185 records in each of 52 branches of a certain chain store. What was the total number of sales of this record throughout this chain store?

$1185 \times 52 = 61\,620$ records

Check by approximation:
$1200 \times 50 = 60\,000$
My answer looks correct!

Finally there will be questions relating to charts, tables, newspaper advertisements, price lists, time-tables etc.

Specimen aural tests

In the next few pages there are specimen aural tests, together with a standard answer sheet. The tests are at the three levels i.e. Foundation Level; Intermediate Level; Higher Level; but the same answer sheet can be used for all levels.

When you sit these aural tests, the instructions from the examination groups may vary slightly but will all follow a similar pattern.

Try all three tests using your parents or friends as 'the teacher'. Fifteen correct answers out of 20 will be a reasonable result for your particular level. You will be given about 30 seconds in which to answer each question after it has been read out.

Instructions (same for all Levels)

You must have nothing on your desk apart from a pen or pencil and your answer sheet.

Each question will be read twice and you will then have a reasonable time in which to write your answer. Write your answer clearly in the space provided on your answer sheet. If you need to jot anything down, then do so on the answer sheet. The test is starting now.

Specimen aural test – foundation level

1 What is the cost of six cakes at twenty pence each?

2 The attendance at a football match was fifteen thousand and ninety-five. Write this attendance in figures.

3 If you spend thirty-seven pence and pay with a pound coin, what change do you want?

4 My lunch must be ready at 12.30 p.m. It takes a hour and a quarter to cook. When should I start cooking it?

5 My share of a pools win was one hundred and fifty pounds. If five of us had equal shares, how much did we win altogether?

6 I think of a number and double it. I subtract three and my answer is fifteen. What was the number I thought of?

7 It took me twelve minutes to drive ten miles on the motorway at a constant speed. How long would it take me to drive a further fifteen miles at the same constant speed?

8 How many fourteen pence stamps can I buy for a pound?

9 What number is five more than nine hundred and ninety six?

10 What is the arithmetic mean or average of the following numbers?

seven; eight; nine; ten; eleven

From now until the end of the test you will need the information on the *answer sheet*, which you will find on pages 80-82.

11 How much would be charged in total for cleaning a jacket and a skirt?

12 The rectangle shown is drawn to scale. Approximately how long is it?

13 Estimate, in degrees, the size of the angle *b*.

14 How much would a family of two adults and one child have to pay to sit in the Stalls?

15 What will be the final monthly repayment if I borrow one thousand pounds repayable over eighteen months?

16 What is the cost of a Sapele door that measures six feet six inches by two feet three inches?

17 I need to be at work at Hamil Road in Burlsem by seven a.m. If I live at the end of Oxford Road, what time is the bus that I have to catch?

18 I have a video cassette tape that lasts for three hours. If I record 'Good Morning Britain' how many minutes do I have left on the tape?

19 What was the difference in temperature in degrees Celsius between Belfast and Manchester?

20 For how many hours and minutes does a car need to travel with lights on in Edinburgh?

Specimen aural test – intermediate level

1 What is the cost of four LPs at five pounds ninety-nine pence each?

2 Write the number seventeen thousand in standard form.

3 Joe pays one pound seventeen pence for two pencils and a pen. If the pen costs eighty-nine pence, what was the cost of each pencil?

4 My dog eats three quarters of a tin of dog food every day. How many tins do I need to buy to last him a whole week?

5 What is thirty per cent of two pounds fifty?

6 What is the name of a triangle with two equal sides?

7 A train travels at a constant speed of ninety miles per hour. How long will it take the train to travel a distance of fifteen miles?

8 In 1988 the first day of May was a Sunday. What day of the week was the twentieth of May?

9 What number is fifteen less than ten thousand and four?

10 I am facing north-east. If I turn through one hundred and thirty-five degrees in a clockwise direction, which direction will I end up facing?

From now until the end of the test you will need the information on the *answer sheet*, which is on pages 80-82.

11 How much would be charged in total for cleaning two blouses and a coat?

12 The rectangle shown is drawn to scale. Estimate the perimeter of the rectangle.

13 Estimate in degrees the size of the angle *a*.

14 How much will one adult and three children pay in total to sit in the Back Circle?

15 What will the normal monthly repayments be if I borrow one thousand pounds repayable over thirty months?

16 What is the total cost of two Plywood doors that measure six feet six inches by two feet?

17 How long does it take, in minutes, for the six-thirty bus from Burslem to do the round trip?

18 What fraction of an hour does 'The Time . . . The Place' last?

19 What is the Celsius equivalent to sixty-six degrees Fahrenheit?

20 For how many hours and minutes does a car need to travel with lights on in Manchester?

Specimen aural test – higher level

1 An article is marked at a sale price of sixty pounds after a reduction of thirty-three and a third per cent. What was the original price?

2 To the nearest whole number, what is the length of a diagonal of a square of side seven centimetres?

3 Pete pays three pounds fifty pence for four pies and a pastie. If the pies are eighty pence each, how much is the pastie?

4 I need two hundred candles for a party. If they are sold in boxes of two dozen, how many boxes do I need to buy?

5 The cost of a meal is six pounds twenty plus VAT charged at fifteen per cent. How much VAT is paid?

6 How many prime numbers are there between twenty and forty?

7 I run twice as fast as I jog and I jog three times as fast as I walk. How long will it take me to run as far as I could walk in thirty minutes?

8 What is the name of the quadrilateral that has only one pair of parallel sides?

9 I measured the length of a line to be twenty-one centimetres. In actual fact it was only twenty centimetres long. What was my percentage error?

10 What is the total surface area of a cuboid measuring two centimetres by four centimetres by five centimetres?

From now until the end of the test you will need the information on the *answer sheet* on pages 80-82.

11 What change from a five pound note will I get if I have a skirt and a pair of trousers cleaned?

12 The rectangle shown is drawn to scale. Estimate its area.

13 Estimate in degrees the size of angle *c*.

14 What change will I get from a twenty pound note if I take my wife and three children to sit in the Front Circle?

15 If I borrow one thousand pounds repayable over eighteen months, what will be my monthly repayments total in the first six months of repayment?

16 What will be the total cost of a Hardboard door and a Teak door, both measuring six feet six by two feet three?

17 The bus driver who drives the six a.m. bus on the round trip also has to do the next trip after he gets back to Burslem. How long does he have to wait before starting the next round trip?

18 I have one hour and twenty minutes free on my video tape. How much will remain free if I record the programmes 'Password' and 'Allsorts'?

19 What was the modal temperature in degrees Celsius at midday yesterday for all the places listed in the table?

20 What was the total time for which vehicle lights were not needed in Bristol for this particular day?

Aural test – answer sheet

1 £

2

3 pence

4

5

6

7 minutes

8

9

10

11 £ ____________

Vale Dry Cleaners

Dress	£2.50	Coat	£2.95
Blouse	£1.25	Jacket	£2.10
Trousers	£1.65	Skirt	£1.45

12 ________

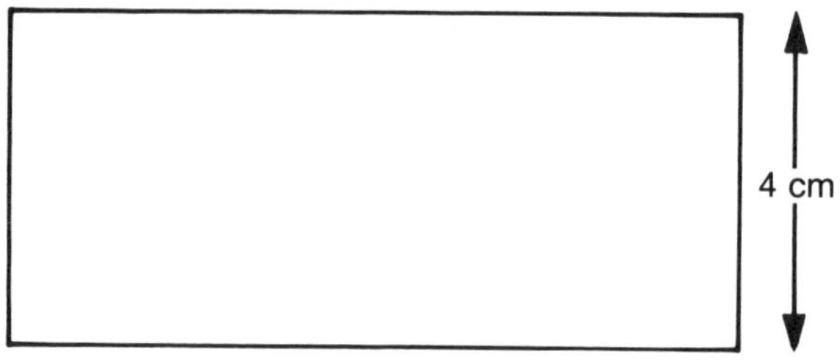

13 ______°

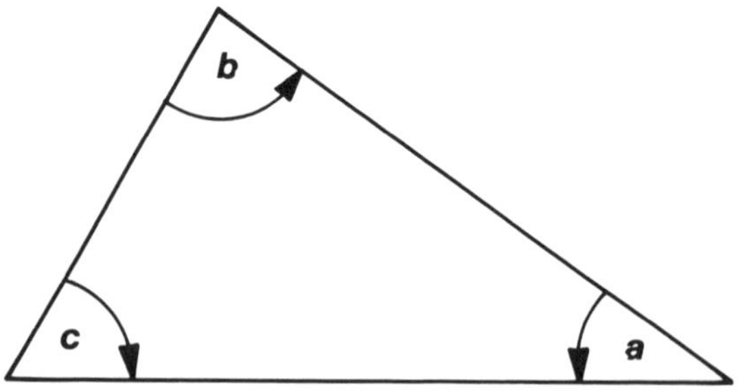

14 £ __________

Prices of Admission

Front Circle	£4.00
Back Circle	£3.20
Stalls	£2.80

Children – Half Price

15 £ ____________

Personal Loans

Supplement to the Personal Loans Leaflet.

The amount which can be borrowed is governed by the ability of the borrower to repay.

The interest charge at 9 per cent per annum is added to the loan and the total amount repaid by equal monthly instalments over the agreed period of up to 5 years.

Representative examples for information purposes.
£1000 Personal Loan at a flat rate of 9 per cent per annum.

	£ 12 months	£ 18 months	£ 24 months	£ 30 months
Amount of loan	1000	1000	1000	1000
Total interest	90·00	135·00		225·00
Arrangement fee*	10·00	10·00		10·00
Total charge for credit	100·00	145·00		235·00
Total amount payable	1100·00	1145·00		1235·00
Normal monthly repayments	90·83	63·06		40·83
Final monthly repayments	90·87	62·98		40·93
Annual Percentage Rate of charge	**19·7**	**19·2**	**18.9**	**18·6**

16 £ ______

PLY FLUSH
DOORS

		Hardboard	H/B primed	Plywood	Colonial H.B.	Sapele	Teak
6′ 8″ x 2′ 8″	2032 x 813mm	14.77	14.16	21.27		30.21	
6′ 6″ x 2′ 9″	1981 x 838mm	14.77	14.16	21.27		30.21	
6′ 6″ x 2′ 6″	1981 x 762mm	14.05	14.11	20.38	41.52	29.05	44.95
6′ 6″ x 2′ 3″	1981 x 686mm	13.95	14.02	20.00	41.00	28.85	44.73
6′ 6″ x 2′ 0″	1981 x 610mm	13.86	13.93	19.61		28.54	44.41
6′ 6″ x 1′ 9″	1981 x 533mm	13.77	13.84	19.60		28.36	
6′ 6″ x 1′ 6″	1981 x 457mm	13.66	13.73	19.28		28.13	

17 ____________

BURSLEM · OXFORD · CHELL HEATH · BURSLEM
BURSLEM · CHELL HEATH · OXFORD · BURSLEM

Mondays to Fridays

BURSLEM (Hamil Road)	0530	0600	0615	0630	0645	0655	0705	0715	0725
Chell Heath (Knave of Clubs)	......	0610	......	0640	......	0705	......	0725	......
Chell (Roundabout)	0537	......	0622	......	0652	......	0712	......	0732
Oxford (End of Oxford Road)	0540	0613	0625	0643	0655	0708	0715	0728	0735
Chell (Roundabout)	......	0616	......	0646	......	0711	......	0731	......
Chell Heath (Knave of Clubs)	0543	......	0628	......	0658	......	0718	......	0738
BURSLEM (Hamil Road)	0553	0623	0638	0653	0708	0718	0728	0738	0748

18 ____________

6.00 TV-am begins with **The Morning Programme** introduced by Richard Keys; **7.00 Good Morning Britain** presented by Mike Morris and Richard Keys.
9.25 Thames news.
9.30 Password. Word association game presented by Gordon Burns. The celebrity guests are Adrian Walsh and Leni Harper **10.00 Santa Barbara 10.25 News** headlines
10.30 The Time . . . The Place . . . Mike Scott chairs a discussion on a topical subject **11.10 Allsorts 11.25 Thames news** headlines.

19 ______°C

YESTERDAY

Temperatures at midday yesterday: c, cloud; f, fair; r, rain; s, sun

	C	F			C	F	
Belfast	13	55	r	**Guernsey**	19	66	s
B'rmgham	21	70	f	**Inverness**	15	59	r
Blackpool	17	63	c	**Jersey**	23	73	s
Bristol	19	66	c	**London**	23	73	f
Cardiff	18	64	f	**M'nchester**	19	66	c
Edinburgh	17	63	c	**Newcastle**	18	64	c
Glasgow	15	59	d	**R'nldsway**	14	57	c

Sun rises: 4.43 am
Sun sets: 9.22 pm
Moon sets 12.47 am
Moon rises 1.14 pm

First Quarter 11.23 am

LIGHTING-UP TIME

London 9.52 pm to 4.14 am
Bristol 10.01 pm to 4.24 am
Edinburgh 10.33 pm to 3.57 am
Manchester 10.12 pm to 4.10 am
Penzance 10.06 pm to 4.43 am

20 ______ hours ______ minutes

Answers – specimen aural tests

	Foundation level	Intermediate level	Higher level
1	£1.20	£23.96	£90.00
2	15 095	1.7×10^4	10 cm
3	63p	14p	30p
4	11.15 a.m.	6 tins	9 boxes
5	£750.00	75p	93p
6	9	isosceles	4 prime numbers
7	18 minutes	10 minutes	5 minutes
8	7 stamps	Friday	trapezium
9	1001	9989	5% error
10	9	south	76 cm^2
11	£3.55	£5.45	£1.90
12	accept 8-10 cm	accept 24-28 cm	accept 32-40 cm^2
13	accept 80°-89°	accept 30°-40°	accept 55°-65°
14	£7.00	£8.00	£6.00
15	£62.98	£40.83	£378.36
16	£28.85	£39.22	£58.68
17	0643	23 minutes	7 minutes
18	35 minutes	$\frac{2}{3}$ hour	35 minutes
19	6°C	19°C	19°C
20	5 hours 24 minutes	5 hours 58 minutes	17 hours 37 minutes

'You should be aiming to get about 15 correct out of 20 at the level you are being entered for.'

You should note that these specimen aural tests do not cover everything that may be asked. Some questions may also involve different aspects of

- metric units
- estimation of length
- rounding up or down etc.

SECTION SEVEN

Answers and summaries

Answers to Section Two – Problem Solving

Problem 3 Diagonals across rectangles

(a) Rectangle measuring 7 cm by 3 cm.
Diagonal passes through (i.e. cuts) 9 squares.

(b)

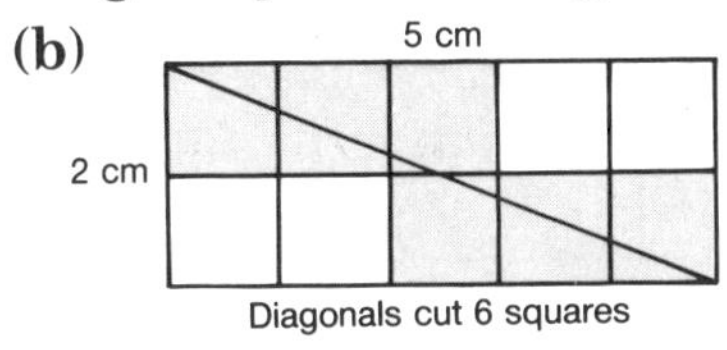

Diagonals cut 6 squares

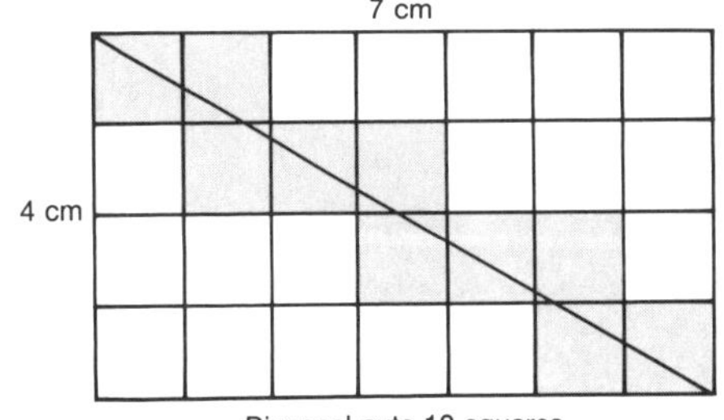

Diagonal cuts 10 squares

Continue for different rectangles and make a table of the results.

Length	Width	Squares cut by diagonal
5	2	6
7	4	10
7	3	9
7	2	8
9	2	10
9	4	12
9	5	13

Length	Width	Squares cut by diagonal
6	3	6
9	3	9
6	2	6
6	4	8
6	6	6
8	4	8
8	6	12

You should look at each pair of lengths and widths and find **common factors**. You will then need the **Highest Common Factor (HCF)** for each pair.

The table has been conveniently divided so that the HCFs for the pairs in the right-hand part differ, whilst for the left-hand part, the only HCF is 1.

The number of squares cut by the diagonal is

$$\text{Length} + \text{Width} - (\text{HCF of Length and Width})$$

So, for those rectangles on the left-hand part of the table; the number of squares cut by the diagonal is

$$\text{Length} + \text{Width} - 1$$

Problem 4 Airport control

Two airports linked would have 2 flight paths.

Three airports linked would have 6 flight paths.

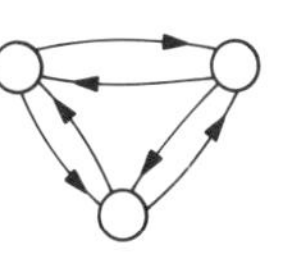

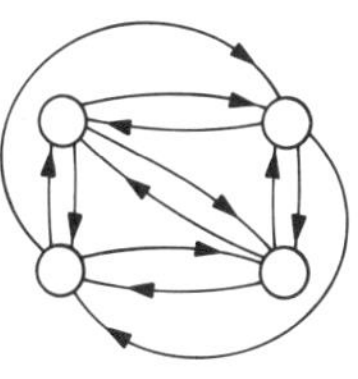

Four airports linked would have 12 flight paths.

Number of airports	2	3	4	5	6	7
Number of flight paths	2	6	12	20	30	42

From the table, generalizing:

$N = A(A-1)$ (where N = number of flight paths
or $N = A^2 - A$ and A = number of airports)

If $A = 200$, then $N = 200 \times 199 = 39\,800$

Therefore, there are 39 800 flight paths linking the 200 international airports.

Problem 5 Blue cubes

Cube dimensions (cm × cm × cm)	Number of faces painted blue 0	1	2	3	4	5	6	Number of small cubes
$2 \times 2 \times 2$	0	0	0	8	0	0	0	8
$3 \times 3 \times 3$	1	6	12	8	0	0	0	27
$4 \times 4 \times 4$	8	24	24	8	0	0	0	64
$5 \times 5 \times 5$	27	54	36	8	0	0	0	125
$8 \times 8 \times 8$	216	216	72	8	0	0	0	512
$n \times n \times n$	$(n-2)^3$	$6(n-2)^2$	$12(n-2)$	8	0	0	0	n^3

Problem 6 Matchsticks in squares

Stage number	Number of matches used	Extra matches needed
1	4	4
2	12	8
3	24	12
4	40	16
5	60	20
6	84	24
20	840	80
n	$2 \times n \times (n+1)$ $= 2n(n+1)$	$4 \times n = 4n$

Problem 7 Timetabling

Allocation of classes:

Mrs Ashcroft	1A	4D	1G	2D	3A	= 20 lessons
Mrs Broad	1B	4E	5A	2E	3B	= 21 lessons
Mr Hill	1C	4F	5B	2F	3C	= 21 lessons
Mr Lewis	3D	2A	1D	5C	4A	= 21 lessons
Mrs Robson	3E	2B	1E	5D	4B	= 21 lessons
Mr Williams	3F	2C	1F	6L	4C	= 21 lessons

(Note that there are other solutions to this problem.)

Problem 8 The Easter Egg machine

(a) 26 400 eggs will be produced over a 50-hour period when the machine is serviced after every 20 hours of production.

(b)

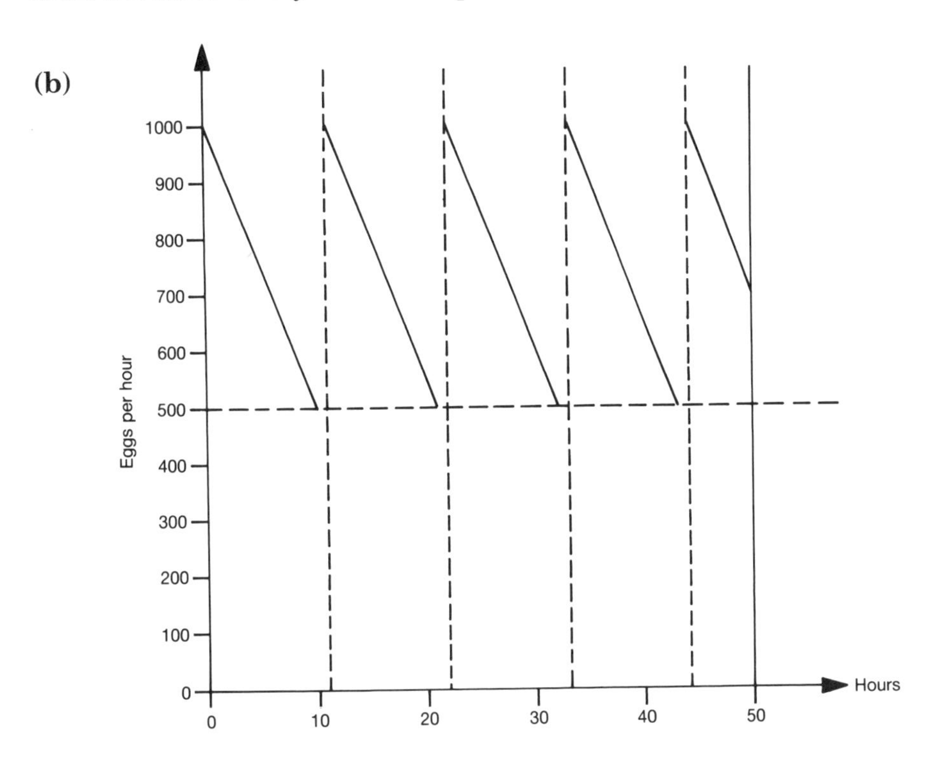

Eggs produced in 10 hours = $750 \times 10 = 7500$ eggs.
In 44 hours the machine will produce $7500 \times 4 = 30\ 000$ eggs.
In the last six hours the machine will produce $850 \times 6 = 5100$ eggs.
Total produced over 50 hours will be 35 100 eggs.

(c)

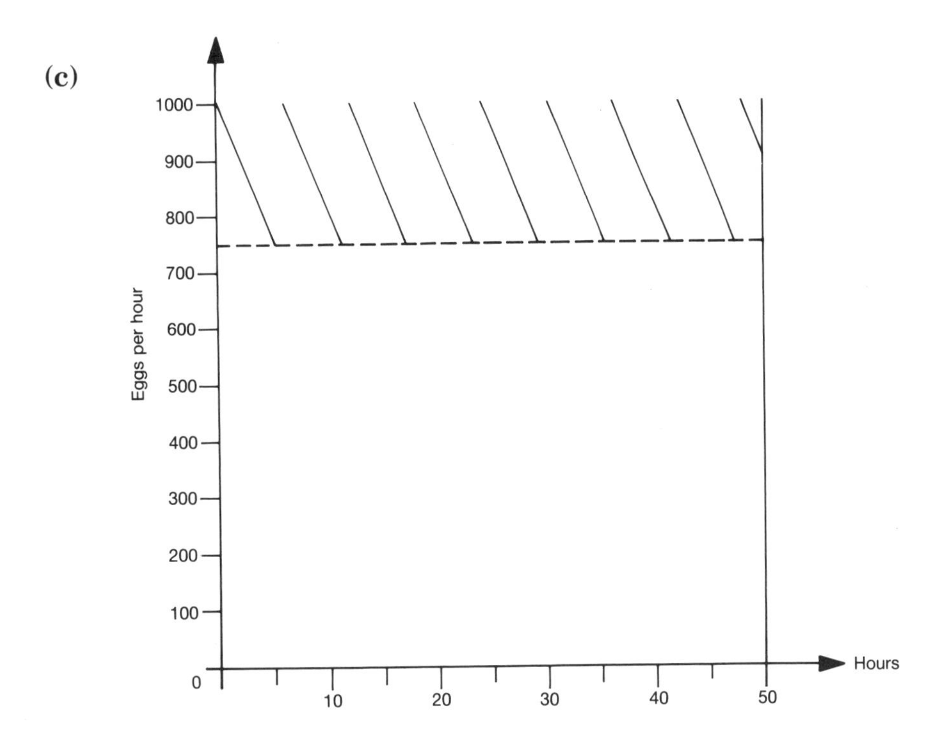

The machine should be serviced every 5 to $5\frac{1}{2}$ hours of production for the greatest output of eggs over a 50-hour period.

Problem 9 Passageways

This question was set by LEAG. The following guidelines are entirely the responsibility of the author.

1 (a) Disruption for seat H is 7.
Disruption for the row is 45 (i.e. $1+2+3+4+5+6+7+8+9$).

(b) Disruption for the row is 24.

(c) Disruption will be at a minimum if passageway is:

J I H G F ↓ E D C B A

(d) Minimum disruption for the Broadway cinema (12 seats) is:

L K J I H G ↓ F E D C B A

Minimum disruption for the Capital cinema (15 seats) is:

O N M L K J I H ↓ G F E D C B A
(or O N M L K J I ↓ H G F E D C B A)

2 (a) Minimum disruption for the Dominion cinema (12 seats) is:

L K J ↓ I H G F E D ↓ C B A

(b) Minimum disruption for the Empire cinema (14 seats) is:

N M L K ↓ J I H G F E ↓ D C B A

(c) Minimum disruption for the Flicks cinema (15 seats) is:

O N M L ↓ K J I H G F E ↓ D C B A

(d) Minimum disruption for the Gaumont cinema (16 seats) is:

P O N M ↓ L K J I H G F E ↓ D C B A

3 (a) Minimum disruption for the Hollywood cinema (12 seats) is:

L K ↓ J I H G ↓ F E D C ↓ B A

(b) Minimum disruption for the Imperial cinema (13 seats) is:

M L K ↓ J I H G ↓ F E D C ↓ B A

Minimum disruption for the Jupiter cinema (14 seats) is:

N M L ↓ K J I H ↓ G F E D ↓ C B A

Minimum disruption for the Kings cinema (15 seats) is:

O N M ↓ L K J I H ↓ G F E D ↓ C B A

Minimum disruption for the La Scala cinema (32 seats) is:

1 2 3 4 5 6 ↓ 7 8 9 10 11 12 13 14 15 16 ↓ 17 18 19 20 21 22 23 24 25 26 ↓ 27 28 29 30 31 32

(These seats are shown 'smaller' so they will fit into one line.)

Note In some of the above, there is more than one best solution although generally, here, only one solution has been given.

As an extension, you may like to look into this.

Summaries to Investigations

Investigation 3 Consecutive sums

$1 = 0 + 1$
$3 = 1 + 2$
$5 = 2 + 3$
$6 = \quad 1 + 2 + 3$
$7 = 3 + 4$
$9 = 4 + 5; \quad 2 + 3 + 4$
$10 = \quad 1 + 2 + 3 + 4$

$11 = 5 + 6$
$12 =$ $3 + 4 + 5$
$13 = 6 + 7$
$14 =$ $2 + 3 + 4 + 5$
$15 = 7 + 8$; $4 + 5 + 6$; ; $1 + 2 + 3 + 4 + 5$
$17 = 8 + 9$

(a) Numbers that cannot be written as the sum of consecutive numbers are: 2 ; 4 ; 8 ; 16 ; Conjecture – powers of 2.

(b) Numbers that can be written as a sum of consecutive numbers in more than one way are: 9 ; 15 ; 18 ; 21 ; 25 ; 27 ; 30

(c) Yes. All numbers except the powers of 2 can be split into consecutive sums.

(d) Yes:
two consecutive numbers = 1, 3, 5, 7, 9, ... all odd numbers
three consecutive numbers = 6, 9, 12, 15, ... add 3 each time
four consecutive numbers = 10, 14, 18, 22, ... add 4 each time
five consecutive numbers = 15, 20, 25, 30, ... add 5 each time
six consecutive numbers = 21, 27, 33, 39, ... add 6 each time
and so on. . . .

(e) Numbers made up of consecutive odd numbers:
$4 = 1 + 3$
$8 = 3 + 5$
$9 =$ $1 + 3 + 5$
$12 = 5 + 7$
$15 =$ $3 + 5 + 7$
$16 = 7 + 9$; ; $1 + 3 + 5 + 7$

Repeat previous investigation.

Numbers made up of consecutive even numbers:
$6 = 2 + 4$
$10 = 4 + 6$
$12 =$ $2 + 4 + 6$
$14 = 6 + 8$ Repeat as before.

Investigation 4 Surrounded by the 'blue tide'

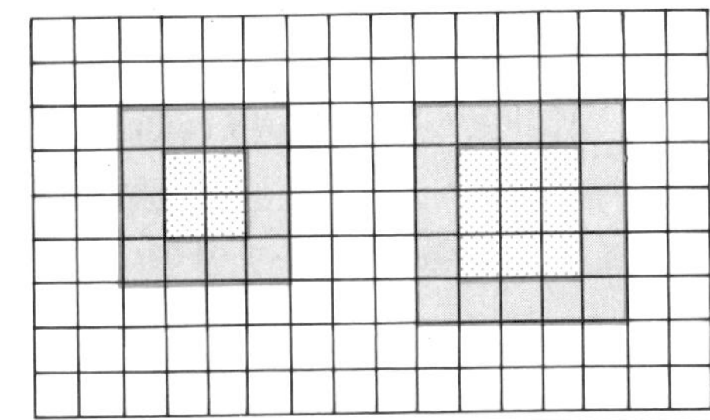

Dimensions (cm)	3×3	4×4	5×5	6×6	7×7
Area (cm^2)	9	16	25	36	49
Perimeter (cm)	12	16	20	24	28
Number of blue squares	8	12	16	20	24
Number of black squares	1	4	9	16	25

If length of square $= x$ then:

area $= x^2$ perimeter $= 4x$ Blue squares $= 4x - 4$
Black squares $= (x - 2)^2$

Extension: try rectangles.

Investigation 5 Crossing lines and regions

Two crossing lines

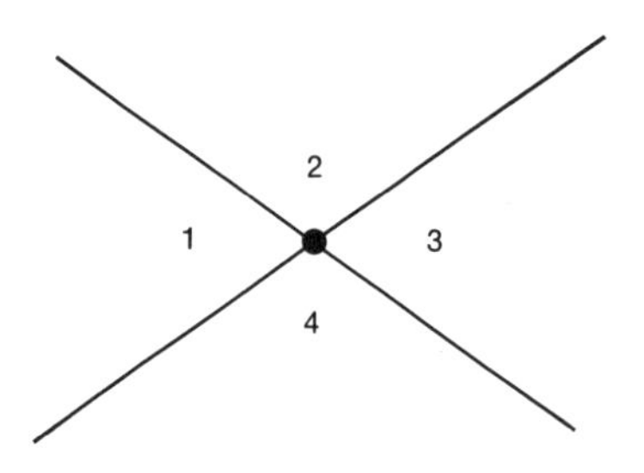

Three crossing lines

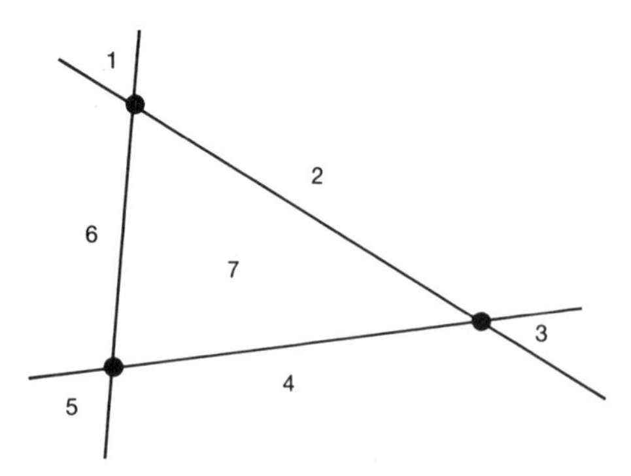

Four crossing lines

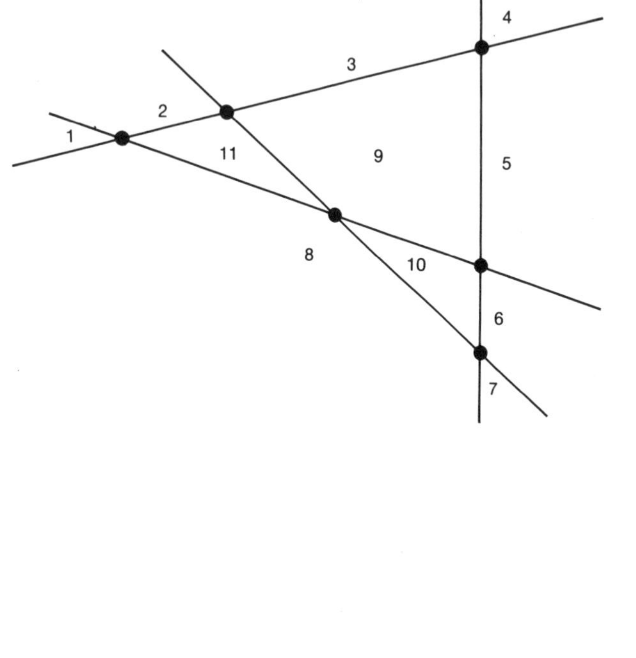

Five crossing lines

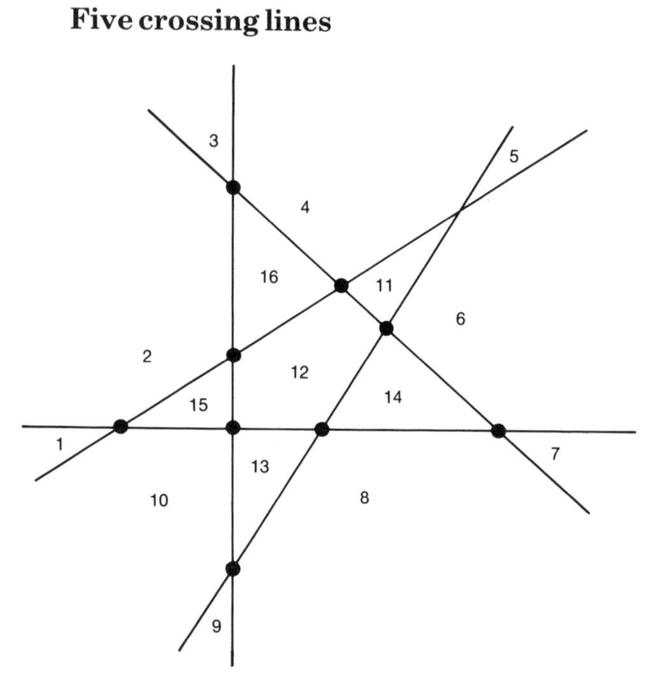

Repeat for six crossing straight lines, then seven crossing straight lines. Make out a table of results.

Number of crossing lines	2	3	4	5	6	7
Number of crossing points	1	3	6	10		
Number of open regions	4	6	8	10		
Number of closed regions	0	1	3	6		
Total number of regions	4	7	11	16		

Look for patterns; make predictions and test them; use algebra to show the relationships.

In general, if you have N crossing straight lines, then:

Number of crossing points $= \frac{1}{2}N(N-1)$

Number of open regions $= 2N$

Number of closed regions $= \frac{1}{2}(N-1)(N-2)$

Total number of regions $= 2N + \frac{1}{2}(N-1)(N-2)$
$= \frac{1}{2}(N^2 + N + 2)$

Extension: consider circles overlapping.

Investigation 6 Changing places – 'Frogs'

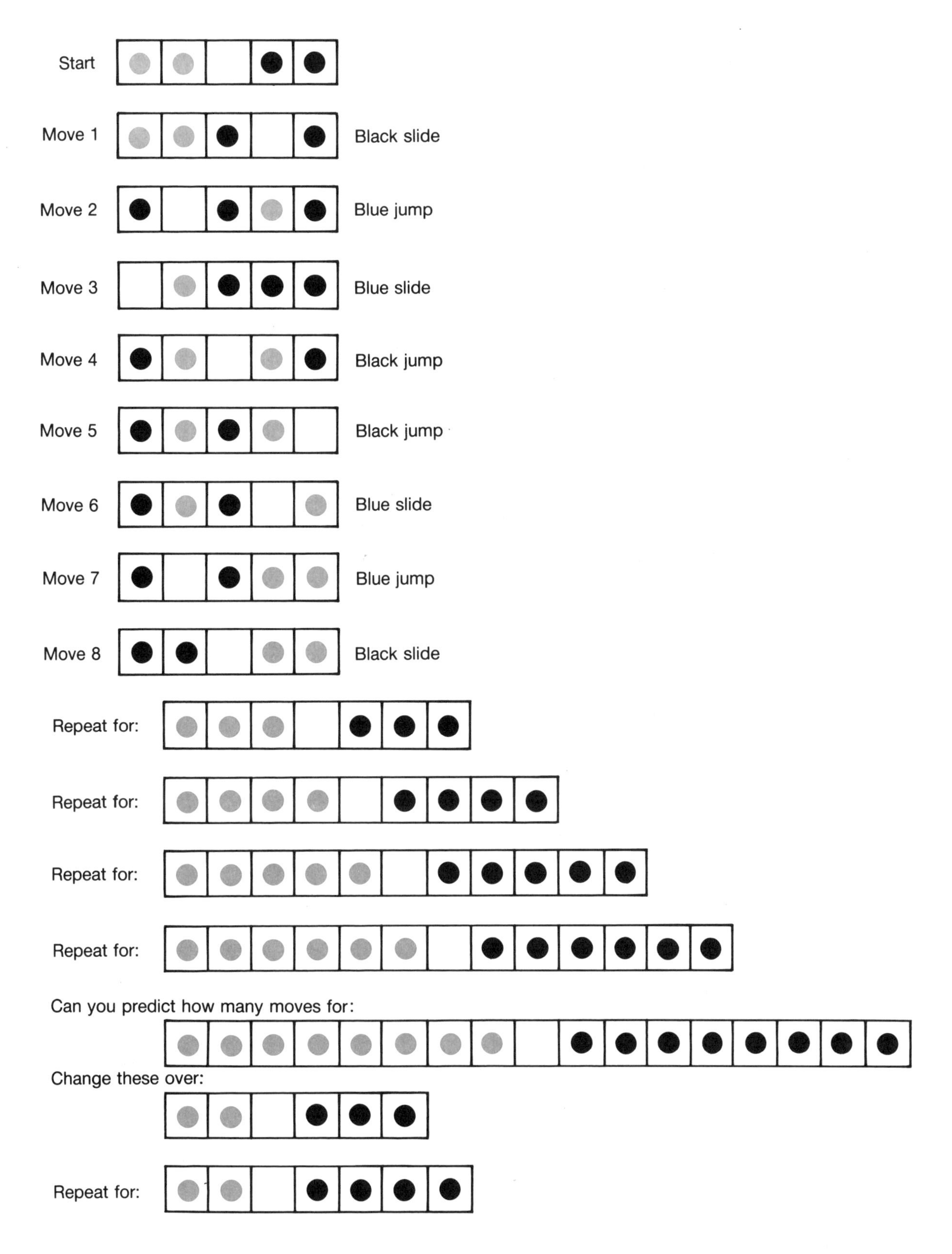

Discuss your results.

Keep going along this path. Can you use the computer? (There are plenty of programs for this investigation.)

Investigation 7 Palindromic numbers

Try any number less than 100:

58	
85	
143	(not palindromic)
341	
484	palindromic after two stages)

59	
95	
154	(not palindromic)
451	
605	(not palindromic)
506	
1111	palindromic after three stages)

Notice that if 59 becomes palindromic after three stages, then so too does 95, i.e. its reverse.

1 2 3 4 5 6 7 8 9 11 22 33 44 55 66 77 88 99
are palidromic as they are, i.e. after 0 stages.

19 28 37 39 46 48 49 57 58 64 67 73 75 76 82 84 85 91 93 94
are palindromic after two stages.

59 68 86 95 are palindromic after three stages.

69 78 87 96 are palindromic after four stages.

79 97 are palindromic after six stages.

Most of the others are palindromic after one stage – in fact all, except two of them.

Investigate these two numbers.

Find exactly how many palindromic numbers there are between 100 and 1000,

e.g. 131 878 626 505 989 etc.

Extensions:

Are there any palindromic **square** numbers?
Are there any palindromic **prime** numbers?
Are there any palindromic **cube** numbers?

Appendix

Comparison of groups' assessment schemes

	LEAG	MEG
Percentage marks *Exam* *Course* *Aural*	70% 25% 5%	75% (70%) 25% (30%) — (including 6% mental)
Examination papers	Each student sits two papers from Papers 1, 2, 3 & 4. Level X: Papers 1 & 2 Grades E-G Level Y: Papers 2 & 3 Grades C-F Level Z: Papers 3 & 4 Grades A-D	Each student sits two papers. Foundation level: Grades E-G (Papers 1 & 4) Intermediate level: Grades C-F (Papers 2 & 5) Higher: Grades A-D (Papers 3 & 6)
Coursework assignments	Students will complete: Five investigations. At least one, but not more than two, from each of the following areas: * Pure investigation * Problems * Practical work	Five assignments, one from each of the following: * Practical geometry * Everyday applications * Statistics/Probability * Investigations (1990) * Centre approved topic (one investigation plus two from the other three categories – not centre approved topics) (1991)
Group's guidance on choice of task	Seven coursework tasks/investigation will be set by the Group to choose from. Candidates will be able to do investigations of their own choice provided they fall into the three stated categories.	Suggested titles are given but candidates are encouraged to select topics outside this list.
Time allowed	Some assignments will be done under timed examination conditions.	2-3 weeks work on each assignment.
Allocation of marks	Each task will be marked as follows: * Correct strategy 7 marks * Implementation and reasoning 7 marks * Interpretation and communcation 6 marks Total 20 marks × 5 100 marks divide by 4 for maximum 25%	For each assignment there are 4 marks for each of the following: * Overall design/strategy * Mathematical content * Accuracy * Presentation/clarity of argument * Controlled element (replaced by oral assessment which will take the form of continuous assessment of coursework, orally, in 1991)
General comments	Three of the tasks will be suitable for all levels. The other two will be chosen according to which level you have been entered for.	Each of the assignments will be assessed by your teacher. Two of them must be done in the final 12 months of the course.
	1990	1990 (1991)

At the time of writing, all information about syllabus requirements was correct. However, because syllabus requirements change, you must **always** check with your examination group or your teacher before you start any coursework, to make sure that you are doing what is required.

	NEA	NISEC
Percentage marks *Exam* *Course* *Aural*	75% 25%	70% 20% 10%
Examination papers	Each student sits two papers from Papers 1, 2, 3, and 4. Level P: Papers 1 & 2 Grades E-G Level Q: Papers 2 & 3 Grades C-F Level R: Papers 3 & 4 Grades A-D	Each student sits three different pairs of papers. Basic Level: Grades E-G Intermediate Level: Grades C-F High Level: Grades A-D
Coursework assignments	Students will complete: Teachers to design coursework together with pupil } Syllabuses A & B	Four assignments. At least one from the two areas: *Practical geometry/Measurement/Everyday application of mathematics/Statistics *Pure mathematics investigation
Group's guidance on choice of task	A list of possible titles is given but teachers are free to devise their own assignments.	Assignments will be devised by the class teacher.
Time allowed Allocation of marks	The length of a piece of work may vary from a few minutes to one done over several weeks.	Each assignment could take two to three weeks.
Allocation of marks	Practical work 30 marks (max.) Investigation 40 marks Assimilation 30 marks Total 100 marks (max.) Divide by 4 for a maximum 25%	*Basic Inter. High* Understanding/Plan 3 5 7 marks Examination of task 3 5 7 marks Communication/ Evaluation 3 5 7 marks Maximum 36 60 84 marks Each scaled down to 20%
General comments	Group work is acceptable but evidence of individual candidate's work must be available. A record of assessment of your ability to take part in a mathematical discussion will be kept.	Group assignments are permitted if individual assignments can be identified.
	1990	1990

SEG	SMP Courses	WJEC
50% 40% 10%	75% 20% 5% } This may vary from Group to Group	Approx. 74% Approx. 22% Approx. 4%
:ach student sits two papers from 'apers 1, 2, 3, and 4. evel 1: Papers 1 & 2 Grades E-G evel 2: Papers 2 & 3 Grades C-F evel 3: Papers 3 & 4 Grades A-D	Each student sits two papers from Papers 1, 2, 3, and 4. Foundation Level: Papers 1 & 2 Grades E-G Intermediate Level: Papers 2 & 3 Grades C-F Higher Level: Papers 3 & 4 Grades A-D	Each student sits two papers from Papers 1, 2, 3, and 4. Level 1: Papers 1 & 2 Grades E-G Level 2: Papers 2 & 3 Grades C-F Level 3: Papers 3 & 4 Grades A-D
;tudents will complete: An extended piece of work Two other 'single tasks' or a series of short assignments One aural test	This varies from group to group. MEG students who follow SMP courses can do either (a) one open-ended assignment plus four set assignments **or** (b) two open-ended assignments. NEA students who follow SMP courses complete a coursework folder.	According to the level of entry. Levels 2 & 3: * A practical investigation * A problem-solving investigation Level 1: * An investigation * Three exercises demonstrating practical skills
'he choice is left to the teacher and student ut must include: Data handling Problem solving Interpretations and generalizations Communication of results	In some cases assignments will be set by the Group. Otherwise assignments will be devised by the class teacher.	All investigations and exercises will be set by the Group. Centres may choose their own provided they are of similar standard.
'he extended piece of work is expected to ake up to four weeks.)ther tasks will be shorter.	Some will be timed at 1 hour, done under test conditions. Others could last up to two to three weeks.	Each task should be given two sessions per week over a four week period. Pupils should devote time outside class time.
:xtended piece of work 48 marks (max.) 'wo other tasks 48 marks (max.))ral aspect 24 marks (max.) Total 120 marks 20 marks divide by 3 for maximum 40%	Varies.	Marks are divided amongst: * Understanding 3 marks * Strategy 5 marks * Content and development 6 marks * Communication 7 marks Also there is a continuous assessment: * Understanding 3 marks * Method 3 marks * Conclusion 3 marks
'ou will be tested orally about your ıssignments, so be prepared to answer ıuestions about them in interviews.		All assignments must be completed by 30 April in the year of the exam. There is continuous assessment, including oral assessment, during the four week period.

1988 1988

Examination groups: addresses

LEAG – London and East Anglian Group

London	University of London Schools Examinations Board Stewart House, 32 Russell Square, London WC1B 5DN
LREB	London Regional Examinations Board Lyon House, 104 Wandsworth High Street, London SW18 4LF
EAEB	East Anglian Examinations Board The Lindens, Lexden Road, Colchester, Essex CO3 3RL (0206 549595)

MEG – Midlands Examining Group

Cambridge	University of Cambridge Local Examinations Syndicate Syndicate Buildings, 1 Hills Road, Cambridge CB1 2EU (0223 61111)
O & C	Oxford and Cambridge Schools Examinations Board 10 Trumpington Street, Cambridge CB2 1QB and Elsfield Way, Oxford OX2 8EP
SUJB	Southern Universities' Joint Board for School Examinations Cotham Road, Bristol BS6 6DD
WMEB	West Midlands Examinations Board Norfolk House, Smallbrook Queensway, Birmingham B5 4NJ
EMREB	East Midland Regional Examinations Board Robins Wood House, Robins Wood Road, Aspley, Nottingham NG8 3NR

NEA – Northern Examination Association (*write to your local board.*)

JMB	Joint Matriculation Board (061-273 2565) Devas Street, Manchester M15 6EU (*also for centres outside the NEA area*)
ALSEB	Associated Lancashire Schools Examining Board 12 Harter Street, Manchester M1 6HL
NREB	North Regional Examinations Board Wheatfield Road, Westerhope, Newcastle upon Tyne NE5 5JZ
NWREB	North-West Regional Examinations Board Orbit House, Albert Street, Eccles, Manchester M30 0WL
YHREB	Yorkshire and Humberside Regional Examinations Board Harrogate Office — 31–33 Springfield Avenue, Harrogate HG1 2HW Sheffield Office — Scarsdale House, 136 Derbyshire Lane, Sheffield S8 8SE

NISEC – Northern Ireland

NISEC	Northern Ireland Schools Examinations Council Beechill House, 42 Beechill Road, Belfast BT8 4RS (0232 704666)

SEB – Scotland

SEB	Scottish Examinations Board Ironmills Road, Dalkeith, Midlothian EH22 1BR (031-633 6601)

SEG – Southern Examining Group

AEB	The Associated Examining Board Stag Hill House, Guildford, Surrey GU2 5XJ (0483 503123)
Oxford	Oxford Delegacy of Local Examinations Ewert Place, Summertown, Oxford OX2 7BZ
SREB	Southern Regional Examinations Board Eastleigh House, Market Street, Eastleigh, Hampshire SO5 4SW
SEREB	South-East Regional Examinations Board Beloe House, 2–10 Mount Ephraim Road, Tunbridge Wells TN1 1EU
SWEB	South-Western Examinations Board 23–29 Marsh Street, Bristol BS1 4BP

WJEC – Wales

WJEC	Welsh Joint Education Committee 245 Western Avenue, Cardiff CF5 2YX (0222 561231)

(The boards to which you should write are underlined in each case.)

INDEX

100538

THE
HUNTINGDONSHIRE
COLLEGE
LIBRARY